INGRAINED

To Alison and Keith (LH)

To Erik and Freja, who would happily eat pasta every day of the week (JA)

To Mum and Dad with love (AG)

Ingrained
A Human Bio-geography of Wheat

LESLEY HEAD
JENNIFER ATCHISON
ALISON GATES
University of Wollongong, Australia

Routledge
Taylor & Francis Group

LONDON AND NEW YORK

First published 2012 by Ashgate Publishing

2 Park Square, Milton Park, Abingdon, Oxon OX14 4RN
711 Third Avenue, New York, NY 10017, USA

Routledge is an imprint of the Taylor & Francis Group, an informa business

First issued in paperback 2016

British Library Cataloguing in Publication Data
Head, Lesley.
    Ingrained : a human bio-geography of wheat.
    1. Wheat--Australia. 2. Wheat farmers--Australia.
    3. Human-plant relationships--Australia. 4. Ethnobotany--
    Australia.
    I. Title II. Atchison, Jennifer. III. Gates, Alison.
    304.2'7-dc23

Library of Congress Cataloging-in-Publication Data
Head, Lesley.
    Ingrained : a human bio-geography of wheat / by Lesley Head, Jennifer
Atchison and Alison Gates.
      p. cm.
    Includes bibliographical references and index.
    ISBN 978-1-4094-3787-1 (hardback)
    1. Wheat. 2. Human-plant relationships. I. Atchison, Jennifer.
II. Gates, Alison. III. Title.
    SB191.W5H43 2012
    633.1'1--dc23

                          2011048725

ISBN 978-1-4094-3787-1 (hbk)
ISBN 978-1-138-26166-2 (pbk)

Contents

List of Figures

List of Tables

Acknowledgements

We thank all interview participants who gave generously of their time and welcomed us into their homes and workplaces. We are particularly grateful to the farming families, for whom the timing of the project coincided with unusually difficult years. This research and writing was funded by grants from the Australian Research Council to LH (DP0665932 and FL0992397). The Bundanon Trust is thanked for a residency to LH which facilitated completion of several chapters. Phil Holmes provided valuable initial contacts. Katarina Saltzman and Birgitta Olsson helped in the field, and added their insights. Cheryl Jecht and Scott East transcribed the interviews, and Pat Muir assisted with analysis. Austen Pepper undertook spatial and temporal analysis. Sean Geltner and Gordon Cameron at the Australian Bureau of Statistics kindly provided historical agricultural data. Research into flour mills was assisted by a grant from the Royal Australian Historical Society to AG. David Virtue, Paolo Abballe and the late Caroline Mitchell helped survey products at the supermarket. Elyse Stanes and Michael Stevens provided a range of technical support, and Diane Walton editorial support. Figures were prepared by Peter Johnson, David Clifton, Hannah Fullagar and Kate Roggeveen. Many colleagues in the Australian Centre for Cultural Environmental Research (AUSCCER) have helped us refine the ideas in this book, particularly Chris Gibson, Gordon Waitt and Nick Gill. Thinking through plantiness is an ongoing collaborative project with Catherine Phillips. For productive discussions in different contexts we thank Lesley Instone, Tess Lea, Elspeth Probyn, Paul Robbins, David Trigger, and participants in the New Biogeographies symposium in Wollongong, and the Producing Regions workshops run by the Hawke Research Institute in South Australia. For support on the home front, advice and discussion we are grateful to Richard Fullagar, Ivars Reinfelds, and Tim Gates.

Earlier versions of parts of Chapters 5 and 9 were published in Head, L., Atchison, J., Gates, A. and Muir, P. 2011 A fine-grained study of the experience of drought, risk and climate change among Australian wheat farming households. *Annals, Association of American Geographers* 101(5), 1089–1108. An earlier version of parts of Chapter 8 was published as Atchison, J., Head, L. and Gates A. 2010. Wheat as food, wheat as industrial substance: Comparative geographies of transformation and mobility. *Geoforum* 41(2), 236–46.

Acknowledgements

List of Abbreviations

AWB	Australian Wheat Board
BCE	Before the Christian Era
BP	years Before Present
bya	billion years ago
CE	Christian Era
DPI	NSW Department of Primary Industries and Agriculture
GM	Genetically Modified
GR	Green Revolution
mya	million years ago
NSW	New South Wales, Australia
SA	South Australia

Chapter 1
Ingrained

Having picked up this book and turned to this page, you have continued the human relationship with wheat, whose starch has been an important component of paper production since the ancient Egyptians made papyrus. Wheat is embedded in Western consciousness as the biblical staff of life, one of the founding grains of the Neolithic revolution, a marker of colonial productivity and progress in the settlement of new lands, and a globalised commodity. It is embedded in everyday lives, including urban lives, and most of us ate it for breakfast without thinking too much about it, unless that is, we suffer from coeliac disease and need to follow its presence everywhere in order to avoid it.

If you were sleepily reading the cereal or bread packet this morning, a picture of a stylised head of wheat or a golden sheaf may have drawn your attention to the fact that you were eating a plant. That plant's ancestors had evolved its cellulose stalk and starchy seeds many millions of years before your ancestors decided to collect those seeds and grind them. Such a picture, or a claim like '97 per cent whole grain' – possibly even a smiling farmer – subtly connotes a natural product and connects your consumer subconscious to the conditions of its production. Perhaps this book is made from 'Nature's paper', fashioned from wheat straw:

> For centuries the farmer has been using the left-over wheat straw for alternative uses. He has put a roof over his head, using it for thatch. He has fed his cattle with it during the winter months. This is a story of recycling long before the word recycle existed. (www.naturespaper.com.au)

However, if your jam or vitamin tablet this morning contained wheat, as is highly possible, the packaging is unlikely to have drawn attention to it with golden sheaves, since to do so would be to render the industrial nature of their production visible. But many legal jurisdictions would have required the label to tell you this somewhere, in case you are allergic to wheat and need to avoid it. Labelling is less stringent in non-food items but if you fed your dog or cat, washed your hair or took a painkiller, wheat was probably involved. Other parts of your putative breakfast were energised by wheat fed to animals. For example, it took half a kilogram of grain to produce a litre of your breakfast milk or yogurt.

Wheat may have not only helped physically constitute your morning newspaper, but also have been on the front page if drought, flood or plague provided a dramatic hook to bring the difficulties of the farming experience into wider community consciousness, if food prices were on the rise or if currency fluctuations were making life hard for exporters. It would certainly have been in the financial pages, where wheat prices and futures on the Chicago Board of Trade are tracked with as much interest as those of tapis crude and gold. Local farmers and traders may also have been celebrating in these pages if poor harvests in the opposite hemisphere promised them higher prices in the forthcoming season.

However ordinary our interactions are with wheat, they are not trivial. It is immensely significant in the global food supply, being the second largest crop produced and consumed by volume, and supplying 19 per cent of calorific supply in the human world (Mitchelle and Milke 2005). Wheat is the crop with the largest production area and a lower global average yield than corn or rice, making its production more energy intensive. It is the largest volume crop traded on an international scale; in 2008 for example 131 million tonnes at a value of 45 billion US dollars was traded internationally (FAOSTATS 2011).

Wheat is the most significant crop for international food aid, and the most significant crop stored as a buffer against production shortages (Mitchelle and Milke 2005, Barret and Maxwell 2005). In 2006, all combined donor countries contributed over 2 million tonnes of wheat and wheat flour in food aid shipments to all combined recipient countries (FAOSTATS 2011). Wheat has been a major agent of landscape change on most continents.

This ambiguous status of wheat – fluctuating between global dominance and everyday invisibility – characterises its treatment within geography. In becoming domesticated, wheat has, for many geographers, crossed from nature into culture, and surrendered its status as a plant. Although it is a grass, you will not find wheat on most vegetation maps, even in places where it dominates the landscape. Nor, although wheatlands were historically part of a system of colonial dispossession and ecological transformation in the Americas, Australasia, South Africa and elsewhere, will you find it in lists of invasive species or weeds of national significance. It has surrendered its status as an autonomous, albeit backgrounded, plant on the stage of human history, and is mapped instead as an economic or cultural category – a regional wheat belt, a flow of trade, a patented variety.

In this book we take a fresh look at the times and spaces of human-wheat relations that are so easy to take for granted. We ask what might be gained by re-approaching these important relations as ones between humans and plants, as beings who are ingrained in one another's lives. This is a smaller scale of analysis than usually undertaken in relation to wheat geographies, but it is a common approach in ethnobotanical studies of the ways hunter-gatherers say, or suburban gardeners, engage with plants. For most of us in the affluent West there is apparently nothing earthy or animated about our relationship with wheat. It is a hard edged industrial relationship, if we notice it at all. How might we approach an ethnobotany, we wondered, among people whose relationship with the plant is via a picture on the cereal packet?

Methodologically, we draw here on a range of approaches from within human geography, biogeography and archaeology. Our empirical contribution is to focus at selected everyday intersections between people and wheat, understanding that these furl in a much wider network of connections between people, places, plants and times. We approach human-plant relations as networks comprising not only plants and people, but also rain, machines, soil, silos, government policies and financial instruments among other things. Our ethnographic windows are through the lives of diverse wheat people – farmers, truckers, millers, scientists, bakers, bankers, pet food manufacturers, traders, and consumers. We also aim to keep the wheat itself front and centre, drawing inspiration from cultural geographers such as Ian Cook's (2004) 'following' studies, Hitchings and Jones' (2004) discussion of mobile and bodily methods for approaching human-plant encounters, and what anthropologists would call multi-sited ethnography. Unlike many 'following' studies, we tried to do so in a way that did not reify wheat as a pre-constituted 'thing' (Figure 1.1). We paid attention to, on the one hand, its material deconstruction and transformation into things other than wheat, and on the other, its interactions with the bodies of humans and nonhumans. This meant following its invisibility as well as its visibility, via food labels and grading standards. It also meant stopping to be aware of the ways wheat was already present and ingrained in our own lives.

Figure 1.1 Following wheat

We argue that our relationships with wheat are just as intimate as those of our ancestors who gathered starchy plants or ground grain on a daily basis. It is deeply ingrained in the fabric of our lives. Nor has it lost what we will call its plantiness, a concept elaborated in Chapter 2. Plantiness is the assemblage of qualities that makes a plant a plant, and that existed many millions of years before humans evolved. Plantiness still requires sun and rain – just enough, not too much, and at the right times. Most of our farming fieldwork took place in southeastern Australia during the extreme drought years of 2005–07, providing unique insights into the interplay of climate with other factors. Our argument hence connects to the wider issues of human-plant mutuality that are so fundamental to the health of the earth's systems.

The book is structured to provide a more theoretical overview in the first half (Chapters 2–4), and more empirical detail in the second half (Chapters 5–9). However this is not fixed, as part of our argument is that the space/times of wheat are simultaneously local and global, of the present day and carrying the past. That is, we understand our ethnographic moments not as small scale contemporary happenings situated within the wider fields of globalisation and evolutionary history; rather each is in a continuing process of becoming that we attempt to convey in the unfolding narrative.

Human-Plant Geographies

Considering that plants are fundamental players in human lives, underpinning our food supply and contributing to the air we breathe, they have received insufficient attention in the social sciences (Hall 2011). In the now well established critique of the nature/culture dualism in Western thought, animal-society relations have received much more scholarly attention than human-plant relations. This is particularly puzzling as relationships between plants and people are central to pressing sustainability issues such as biodiversity protection, food production, invasive species management, biofuel production and carbon sequestration. Jones and Cloke commented a decade ago that, while there had been considerable recent interest in animals and society within human geography and anthropology, 'flora ... remains an even more ghost-like presence in contemporary theoretical approaches' (2002: 4). Chapter 2 goes in search of these ghostly flora, pausing at Matthew Hall's recent (2011) work to consider how the backgrounding of plants became so entrenched in Western life and scholarship. We note that plant related studies are flourishing within the more-than-human turn in geography, and argue that more can be done in response to Jones and Cloke's challenge to take plants seriously. This requires tackling the ontological question, *what is a plant?*, and the epistemological questions that flow from that; *how are plant worlds approached in different human framings*? We forward the concept of plantiness to help us advance thinking about what it means to be a plant, how plants act in their worlds, and how we can we better understand our shared worlds. A further challenge is the question of what the relevant unit of study and engagement might be. In contrast to animals, humans are more inclined to think of plants not as individuals but as different types of collectives or assemblages – forests, food, commodities, vegetation communities, habitat, biodiversity and, more recently, carbon storage devices. These differences have consequences for our ethical engagement.

If human geographies have been rather slow to take plants seriously, within biogeography there has until recently been a converse gap on the question of humans. Although biogeography, as part of the natural sciences, would in theory claim a holistic remit that includes humans as part of earth's biota, its usual practice has reinforced humans as different and separate to the rest of nature, with anthropologists or archaeologists more likely to consider humans within an explicitly biogeographical perspective. As Terrell argues, 'the expression "human biogeography" sounds both self-evident and yet peculiar' (2006: 2088). Most physical biogeographers now recognise that the vegetation patterns they are studying reflect both deep time evolutionary pathways and the 'muddy and indecipherable blur' of human influence (Mackey 2008: 392), but 'an outdated view of the world as "natural ecosystems with humans disturbing them"… remains the mainstream view' (Ellis and Ramankutty 2008: 445). Ellis and Ramankutty have recently put forward the concept of anthropogenic biomes to challenge this view, work we also examine in Chapter 2. We review literature that shows the continuation of dualistic frameworks, as well as challenges to them, and also consider the problematic place of agricultural landscapes in separationist understandings of nature. Recent reconceptualisations of domestication in archaeological thought are discussed as a way of developing more-than-human perspectives over deep time.

Our own take on human-plant geographies (expressed in the human bio-geography of the book's title) draws together plantiness, becoming, and webs of relationality and multiple agency. It considers questions of difference, identity and collectivity, since many of our relations with wheat are not with the individual plant itself.

What is Wheat? And How Did it Become Itself?

'Wheat' is a plant collective that defies our efforts at defining and describing its geography in several ways. Its botanical taxonomy is enormously complicated, with all attempts at its systematic classification proving very difficult (Morrison 2001). Wheat is an ethnobotanical term rather than a specific taxonomic identity. It encompasses two genera (*Triticum* and *Aegilops*, Morrison 2001) and approximately 600 species (Cornell and Hoveling 1998). The taxonomy is complicated by the existence of multiple genetic strains, and by its huge genetic diversity, which is about six times larger than the maize genome (Huttner and Debrand 2001), the result of complicated historical and ongoing interbreeding and hybridisation by humans over thousands of years of domestication (Feldman, 2001). Just as other domesticated food plants emerge as 'countless', 'impossible to avoid' (Cook 2004) and 'invisibly ubiquitous' (Whatmore 2002), wheat is – even more so – simultaneously visible and invisible, obvious and hidden, everywhere and nowhere.

Chapter 3 starts to untangle this history by examining how wheat became itself. Although we use four long moments in an historical trajectory, we are interested in emphasising that this was, and continues to be, a messy and contingent process, an experiment that is still in progress. The chapter discusses becoming grass, becoming domesticated, becoming Australian and becoming a global commodity. To emphasise that the process of becoming is a constant one, we draw attention to different ways in which wheat's identity has been stabilised and disrupted. The process of becoming is also and always a spatial process; this chapter uses examples from many different places and scales.

Figure 1.2 Wheat in the paddock

Spaces of Wheat: Shadow Places

Chapter 4 starts with an overview of landscapes of wheat as conventionally understood – where it grows, and why (Figure 1.2). But we aim to go further and disrupt that view, drawing on more dynamic conceptualisations of spatiality in recent human geography and related disciplines. Plumwood (2008: 139) used the idea of 'shadow places' to bring to light the many unrecognised 'places that provide our material and ecological support, most of which, in a global market, are likely to elude our knowledge and responsibility'. Many urban dwellers are groping for some sort of connection to rural landscapes; our contention is that they already exist. Indeed cities cannot have developed without the underpinnings provided by wheat and related crops, and the storable surplus they generated. These crops are still threaded through cities via markets, transport, products and of course consumers. To cross the conceptual boundary between rural and urban spaces is to also challenge the notion that cities are places of pure culture, outside nature. The material connections between cities and their shadow places have been demonstrated by a number of scholars, showing how cities are networked into broader ecosystem processes, such as those that supply water (Gandy 2002, Kaika 2005, Heynen et al. 2006). This is consistent with the idea that cities are just as much ecosystems as remote wilderness areas (Botkin and Beveridge 1997).

Hence the shadow places of wheat extend far beyond the iconic farm; wheat is ingrained into the nooks and crannies of complex globalised lives and bodies (Figure 1.3). Its shadow places include the branded shapes of pet food, vaginal pessaries, Saddam Hussein's Iraq, and financial markets from Chicago to Singapore. And nor do most wheat farms conform to the stereotype of degraded shadow places, remote from the cosmopolitan centre. Many farmers expressed to us their annoyance at the cliché of drought and farming as represented in the metropolitan media; a sad family standing in a cracking clay pan. Rather they wanted to express their business savvy, the normality of drought, and the necessity of living with risk and uncertainty.

Figure 1.3 Wheat and bodies

Chapter 5 examines the seasonal wheat cycle on the farm, considering what it takes to grow a crop in the wheat belt of New South Wales, Australia. We present the growing of wheat as a complex coproduction between human, plant and other bodies. The human bodies, in the drought years of our fieldwork, underwent as many stresses and strains as the wheat itself. The chapter also emphasises multiple temporalities; it juxtaposes cyclic and rhythmic notions of time with interruptions and unpredictability, in order to cut across a simplistic linear reading of the flow of wheat.

As wheat is harvested and leaves the farm, it is no longer constituted as an individual, or even necessarily a collective, plant. This sets up a number of tensions and issues which are explored in the following three chapters. On the face of things the sequence of mobility (Chapter 6), becoming food (Chapter 7) and industrial transformation (Chapter 8) is a linear process through which wheat further loses its identity as a plant, and increasingly becomes a commodity or a collection of chemically interesting starches and proteins. However, we show that wheat retains its plantiness; indeed that is precisely its attraction to the many humans who engage with it in the contexts of these chapters. Just as, following the argument of Chapter 3, wheat has always had multiple identities, so this continues in various forms of transformation. For those who deconstruct wheat and reassemble it into other things in the industrial laboratory, it is the planty qualities of starch chemistry with which they engage. Others who are turning wheat into meat or milk are more interested in the aggregated plantiness of wheat as energy.

Mobility, Friction and Fungibility

Chapter 6 considers the conceptual and physical interplay of lines and boundaries on the ground, using themes of mobility and friction. The archaeological legacy of older wheat infrastructures in country towns – silos, mills, railways – is one important expression of wheaten landscapes. While today they can look less than dynamic, in their heyday they represented movement and flow just as much as storage and sedentism. Wheat has long been more mobile than its non-perishable character would suggest. An important thread through this and the following chapters is the role of quality standards and other forms of measurement in fixing – or unfixing – wheat's identity, to enhance human connections with it and reduce the friction associated with various forms of movement. We trace wheat's integration into a single stream, and then its differentiation into wheats with a range of different identities.

When you spend a lot of time thinking about the infrastructures that move wheat around, it seems somewhat quaint that it is still pinned to the ground for the growth cycle, dependent on rain to not only fall in the right amounts, but at the right time. From the moment of harvest it is on the move. Friction in this flow is expensive – it costs time and money to load and unload trucks, trains and silos. In his book on Chicago, William Cronon (1991) described the way the movement of wheat from the US mid-west gradually became more fluid with changes in technology. This process is intensified – it is literally a flow – in contemporary grain movements. Not only does it flow, but it is monitored and accounted for like the waters of an intensively managed river, or the rivers of global capital. Today, just-in-time processing drives and punctuates its flow even more. Sydney's largest biscuit factory keeps only two days supply of flour.

This mobility is also more complex than we usually think. While railway webs still drain inland areas to coastal ports in a dendritic pattern, the flow of wheat is not only in one direction. It moves around, re-energising other parts of the network, fuelling and feeding as it goes. As encapsulated energy, it can be stored and pooled, moderating seasonality and the variability of drought. Farming households in Western agriculture do not necessarily grow their own food, but buy back supermarket products including bread and pasta. They never know whether their own wheat went to feed the pigs whose chops they are buying.

Food

A finely tuned and 'just in time' food transport and distribution system which 'presents risks of rapid spread of contaminated food and is vulnerable to events such as pandemics' (PMSEIC 2010: 1) is just one of the challenges of global food security. During the period of our research, dramatic increases in the price of food led to riots in many parts of the world. In contributing to these debates, contemporary agri-food geographies aim to work across, and eventually dismantle, a set of binary oppositions that have marked the field: culture and nature; conventional and alternative agriculture; global and local processes; production and consumption; political economy and cultural approaches; foci on materiality and representation (Morgan et al. 2006). Chapter 7 connects to these debates, focusing on wheat as a staple food, via the examples of bread and pasta.

Because food is central to the interactions between human bodies and the nonhuman world, it is not surprising that heated discussions around the concepts and practice of cultures/natures have been central to food geographies over the last decade. Goodman (1999) argued that this field, where nature was 'abstracted from the social domain' (p. 17), was in fact rather late to examine its basis in modernist ontology. He advocated actor-network theory as a potential way forward, an argument challenged by Marsden (2000), who argued that relational approaches, while offering 'a better methodological tool kit' could not sufficiently account for the power differentials evident throughout agricultural networks.

While we are conscious of the political-economic issues that attend wheat as food, we follow here Marsden's suggestion that empirical studies with a 'significantly more micro-sociological stance' (p. 27) provide a way forward. Our ethnographic examples compare and contrast factory-based, large scale production processes (that we call Big Bread and Big Pasta) with artisan processes (Small Bread and Small Pasta), examining how wheat 'becomes food' (Roe 2006a) in each. In these explorations a further set of comparisons emerge; mass and specialist food production, handmade and untouched food, quality as standards and quality as materiality. The everyday experiences of these food producers constitute part of the broader political economy of wheat. In the struggle to make a living by adding economic value to an apparently simple plant, they have to move it around the continent and pull it apart in different ways. Arguably, they have to get further from the plant. At one level this conflicts with the nutritional and health aspirations that our food should be as fresh and unprocessed as possible, but as boutique liquorice maker Nick wryly explained, he would hardly make a living out of putting bowls of flour on the table. This chapter also adds to food geographies that are challenging the simplistic equation of local food with sustainability concerns, by illustrating some of the complexities of moving organic flour around the country.

Transformation

Wheat has been turned into money since our ancestors recognised its starchy, floury and kneadable qualities. The diverse, energetic and functional qualities of wheat plantiness, however, have enabled us to pull it apart in myriad ways. Its distinctive physical characteristics are central to its transformability and contemporary industrial applications. All conventional grain crops are significant sources of carbohydrates (both starch and sugar), but wheat contains considerably more protein than rice, maize, barley or millet (Pomeranz 1988). Although soybeans have a higher protein content, they do not contain comparable quantities of carbohydrates. Wheat can thus become a component of hairspray, paper and milk via contemporary agricultural and industrial modes of production and processing. It moves in nonlinear ways, hides, and in turn transforms other things.

Chapter 8 examines the processes of transformation and in/visibility. Manufacturers, food technologists and industrial scientists value the malleability and unstable identity of wheat. It is easy to hide, break down and reconstitute into different products, from dog food to ice-cream to hairspray. Its capacity to be broken down as different constituent parts and recrafted into other things is fundamental. We follow the wheat as it becomes transformed into different products, both food and non-food.

Chapters 7 and 8 show that food tends to fix the identity; non-food tends to hide, make invisible and disassemble the identity of wheat. These multiple identities and connection to plantiness make the ethical issues complex. Is our responsibility to plants, or to a less essentialised constituency?

Risk, Drought and Climate Change

One of wheat's metaphorical transformations is as wheat futures, leading us to a discussion of risk and climate change. In Chapter 9 we look at how wheat is imagined in different climate change scenarios; this is just one aspect of the immensely complex challenges of the future. Although drought is a regular feature of Australian wheat farming, the long cycle of drought over the last decade or so is unusual within living memory. It provides a research window onto climate change scenarios of more frequent droughts. Just as drought is shown throughout the book to have considerable agency in the different wheat becomings, so it has a potential role in the sociocultural transitions that climate change will require. Yet even in the context of drought, 'climate' and hence 'climate change', is not expressed or experienced 'separately' to anything else. Climate change will have expression in localised and temporally specific weather processes recognisable in the present. And further, it will also have expression in assemblages comprising 'more-than-climate', assemblages which will include fluctuating prices, public discourse and legislation, measurements of new kinds of value (carbon and water footprints) and human and other bodies.

Bodies

In attending to the bodies of both humans and wheat, we are influenced by a generation of feminist scholarship. This helps us draw together apparently disparate themes such as the role of agricultural colonisation in the dispossession of female plant gatherers (Gott 1982, Knobloch 1996), methodologies of corporeality and embodied experience (Probyn 1993, 2000; Longhurst 2001) and the gendered nature of rural life (Alston 1995, Bryant and Pini 2011). So we were attuned to the gendered nature of the contemporary agro-industrial wheat network. Most of the voices in this book are those of men, who dominate the professional networks through which wheat flows. We had to work much harder to find and draw out female voices. In farming households the enterprise as a whole is clearly dependent on female as much as male labour, not only in domestic work but also often in the business aspects that are now so important in maintaining farm viability. However, we usually found both partners more likely to present the man as the farmer, as the one who interacts with wheat. The exception to this was the most financially successful households, who presented themselves as family enterprises in which a range of specialist roles were acknowledged, including those of adult children. In these households the business role of women was very much to the fore, and class differences intersect with gender.

Figure 1.4 Country Women's Association meeting, Temora NSW, 2007

Of course, organisations such as the Country Women's Association (Figure 1.4) have long supported, drawn attention to and validated the female labour that underpins rural life, particularly among older generations. So it is very easy to find embodied connections between women and wheat in the kitchen, where wheat is kneaded, stirred, cooked and eaten (Figure 1.5). In this respect the lives of contemporary urban and rural women in Australia are not that different; the supermarket is the main place they get their wheat, and a key site of ethnobotanical encounter.

It is one thing to recognise that the choice of methods can privilege some human voices over others, and think of ways to deal with this. It is quite another to try and give voice, and body, to wheat itself. We recognise that our rendering is inevitably partial and problematic. The purpose of approaching the wheat story as a human-plant encounter is not to separate out one particular engagement in a much more complex assemblage. Quite the opposite; we are interested in understanding the world in an associative rather than separationist way. However it is our contention that understanding those associations, and fostering healthy collaborations into the future, requires close attention to the distinctive characters and capacities of the actors involved. To provide a deeper understanding of how wheat and humans have become ingrained in each other's lives it is necessary to try and consider each on their own terms.

Figure 1.5 Cringila community day, Wollongong, 2007

Chapter 2
Reapproaching Human-Plant Geographies

In order to meet the book's aim of advancing understanding of human-plant relations using the example of wheat, this chapter draws on recent debates in human geography, biogeography, archaeology and related disciplines. In each of these areas scholars are grappling, in different ways, with how to move beyond the 'stale binaries' (Anderson 2005: 280) of nature and culture, humanity and animality, hunter-gatherer and agriculturalist, ecology and society. Other binaries are not so stale but just as problematic; invasive and native plants, or vegetation community and crop. The point of going beyond binaries is not to get rid of difference, an important comparative tool, but to ensure that it is carefully explicated on the basis of empirical evidence, rather than categorised from above, and that it does not automatically translate into a hierarchy (Plumwood 1993).

Geographers have long been interested in questions of difference between human groups as well as between humans and various other-than-humans. The concept of encounter has emerged in this literature to discuss various ways of approaching difference ethically (Watson 2006, Fincher and Iveson 2008). In a key overview Valentine reflects on 'how we might forge a civic culture out of difference' (2008: 323), emphasising that contact with (human) 'others' does not necessarily translate into respect for difference (Valentine 2008: 325). Valentine searches for the kind of encounters and spaces that produce what she terms 'meaningful contact', contact that does translate into respect for difference. This approach informs our approach to relations between both humans and plants, and between different disciplinary traditions and methodologies.

Our starting point that the human-plant (or indeed the human-animal) difference is the one to be explained, is an ongoing concern within a Western enlightenment ontology (Anderson 2008). (We return below to the question of whether it is a *peculiarly* Western concern.) In his challenging book, *Plants as Persons. A Philosophical Botany*, Matthew Hall (2011: 6) argues that the dominant Western relationships with plants have been characterised by separation, exclusion and hierarchy. He traces the influence of ancient Greek philosophical traditions, and the Judeo-Christian bible, into the botanical sciences. Hall's detailed historical and philosophical treatment of the backgrounding of plants within these traditions helps to understand the gaps, silences and entrenched assumptions within the disciplines of interest here – why flora have remained ghostly in the social sciences, and why humans have been mostly excluded from nature in the ecological sciences and biogeography. Within botany, the discipline which arguably engages most closely with the materiality of plants, there is evidence that scholars have both absorbed the view that plants are a lower form of life, and been challenged by the evidence before them that this is not the case. This interplay between empirical challenge and imposed categories is also seen in the encounter between indigenous peoples and colonising agriculturalists, and in archaeology's debates over the evolution of agriculture and domestication.

The 'establishment of plants as passive creatures with no capacity for intelligence or communication' (Hall 2011: 19) is first seen in the philosophy of Plato and his hierarchical dualism of reason over nature. The purpose of plants in Plato's *Timaeus* 'is not to live and blossom for themselves, but to provide animals and humans with food' (Hall 2011: 22). The separation of plants and animals was continued by Aristotle, who judged plant capacities by what he observed in animals; he defined animals as those who move and plants as those who do not (*Parts of Animals*; see also Thanos (1994)). Hall provides other examples of Aristotle evaluating plants against zoological criteria and thus finding them lacking; they do not process food and do not excrete waste products. 'As the food they needed is presumed to be adequately concocted in the ground, there must be no activity involved' (Hall 2011: 27).

The divide is traced into the biblical creation stories, and the story of the great flood, when plants are not included in the living things that were blotted out from the face of the earth. 'This floating wooden ark made from the parts of nonliving plants and filled with living animals encapsulates the radical ontological divide between plants and animals' (Hall 2011: 59). Many writers have commented on the pervasiveness of agricultural themes within the bible. For Hall, this is evidence that 'across the range of Biblical sources and ecosystems, the human-plant relationships are consistently characterized as instrumental' (2011: 65). For example in Deuteronomy, Canaan the promised land is defined in terms of seven agricultural plants: 'a land of wheat and barley, of vines and fig trees and pomegranates, a land of olive trees and honey [from date palms]' (Deuteronomy 8: 9, cited in Hall 2011: 65). 'Plants are based at the bottom of a hierarchy of the natural world and are excluded from human moral consideration' (Hall 2011: 8).

These hierarchical connections between immobility, passivity and lack of intelligence have become embedded in our language. The verb 'to plant' means to fix something in a particular location. In English, we use the term 'vegetative' to connote stasis, in contrast to the liveliness of the verb 'animate' (Hitchings 2007: 1). But these trends were not inevitable, and do not in fact relate to innate physiological differences between the organisms. Hall also argues that along the way there were systematic detours when alternative perspectives on plants were not chosen. In particular, Aristotle's flawed understanding of plant anatomy was not shared by his student Theophrastus, often known as the Father of Botany. Theophrastus was careful to approach plants in their own terms, not by comparison with zoological criteria. For Theophrastus, plants were 'volitional, minded, intentional creatures that clearly demonstrate their own autonomy and purpose in life' (Hall 2010: 8).

Why then did Western botany absorb more of the Aristotelian than Theophrastian perspective? Hall attributes this to a 'lack of proper engagement with living plants and the muddled copying of previous scholarship' (2011: 39), by later scholars such as Pliny. The works of Theophrastus were lost to European scholarship from around 200–1500 CE, whereas Aristotle's influence flourished, for example in the work of Thomas Aquinas. Hall argues that Aristotle's dogma of hierarchy and domination was retained by pioneers of scientific methodology such as Francis Bacon.

> Bacon buried Aristotle's archaic views on plants deep into scientific thought by establishing the position of exclusion as the status quo. In this respect, Bacon's work marks a key threshold in the development of the perception of plants. In other areas of science, Bacon advocated the empirical experimental method for uncovering truth, but in his understanding of plants, he happily and entirely subscribed to the wayward deductive reasoning of Aristotelianism. Bacon failed to call for experimentation on the nature or faculties of plants as the dogmatic subscription to the inferiority of plants fitted well with his designs for enslaving the natural world. (Hall 2011: 46)

Bacon is just one example drawn on; Hall goes on to discuss the uncritical acceptance of the Aristotelian hierarchy by the botanists Joachim Jung, John Ray and Linnaeus, and the philosophers Descartes and Locke. Aspects of these trends are evident in how human-plant relations have been treated in the disciplinary traditions of cultural geography, biogeography and archaeology. We review aspects of each of their perspectives below, including recent moves to shift the register of discussions. (It is important to emphasise that we are not attempting a comprehensive review; rather, we aim to identify the most interesting and relevant trends within each field.) We then go on to summarise the perspective we are adopting to analyse human-wheat relations, showing how our perspective extends and further develops these themes. Our aim is to extend discussions in the more-than-human vein, 'where a concern for the character and consequence of nonhuman difference has become central' (Lorimer and Davies 2010: 33), to consider in particular the character and consequence of plant difference. A question that has not been clearly articulated in recent work is, what makes a plant a plant? Engaging with this question in this chapter through the concept of plantiness helps us advance thinking about how plants act in their worlds, and how we can better understand our shared worlds.

More-than-human Geographies: In Search of Ghostly Flora

Cultural geographers have been at the forefront of recent social sciences scholarship disrupting the previously unproblematic divide between humans and the rest of the natural world. Influenced by scholars such as Latour (1993), human geographers have argued that we need to engage with a world that is 'more-than-human', using concepts such as networks and hybridity (Whatmore 2002). Of course there are long traditions within cultural geography of studying human relations with plants and animals, for example in the work of Carl Sauer (1952). The key difference with the more-than-human turn is its emphasis on relationality and multiple agency; that categories and configurations of human entanglement with the nonhuman world (for example, domestication) are not pre-existing givens, but become and are worked out in a process of relation. Geographers and others have contested the idea and practice of human exceptionalism, and have used this to rethink human identity and subjectivity (Anderson 1997, 2008, Emel et al. 2002, Haraway 2008).

The other-than-humans receiving most of this attention have been animals. The new animal geographies have tended to be preoccupied with a set of themes that have dominated, and arguably come to stand for, more-than-human geographies. Most significant of these is the question of boundaries between humans and other animals. Methodologically, many animal geographies proceed from encounters with the human, in which the agency of the animal and the permeable boundaries with humanness are acknowledged. The underlying aim (sometimes implied) is to interrogate similarity and difference, closeness and distance (including for example, Hayden Lorimer 2006 on herding animals, Haraway 2008 on dogs and Jamie Lorimer 2010 on elephants). If the human cannot be privileged in quite the same way, there are many challenging and productive discussions to be had around how we develop more ethical relationships with animals (Whatmore and Thorne 1997). And there are spatial implications of our bounding practices; which animals/species are included or excluded from spaces, for example through their designation as wild, feral, companion or pet?

Inspired and also provoked by Haraway's multispecies thinking with a focus on dogs, scholars are attempting to go beyond 'intuitive and benign encounters between stable, coherent, and large mammals' (Lorimer and Davies 2010: 32), to consider viruses, mosquitoes (Beisel 2010), bacteria (Hird 2009), and the indifferent earth itself (Clark 2011). Lulka (2009) has discussed the residual humanism in geography's use of the hybridity concept when 'nonhumans' are lumped as a singular entity. He called for a 'thick hybridity' in which an adequate sense of difference is maintained. Yet even some of these critical cultural geographies have absorbed a taxonomy with assumptions, for example that immobility is a defining criterion of plants (Lulka 2009: 386, Lorimer 2010: 493).

The more-than-human turn in geography has seen a number of plant-related studies in recent years (Whatmore 2002: Chs 5 and 6, Jones and Cloke 2002, Hitchings and Jones 2004, Power 2005, Franklin 2006, Robbins 2007, Head and Atchison 2009, Instone 2010) and we seek to build on that work in this book.

Jones and Cloke (2002: 8) argued that there is 'considerable scope for widening discussions of non-human agency to embrace beings or entities which are more markedly different than animals from the human'. Their examination of trees stands as one of the most substantive works in human-plant geographies to provide a 'detailed and grounded account of the non-human agency of particular beings, things and materials' (Jones and Cloke 2002: 48). Since Jones and Cloke were writing, the agency of plants has been examined and talked about in a variety of ways. Plants are described as using a variety of strategies, from capturing the care and attention of humans through being 'fun and desirable', to adopting a more stoical approach that can involve 'studied indifference to the gardener' (Hitchings 2003: 105–6). Herbs make themselves visible in the forest to facilitate collection by humans (Staddon 2009a), and trees are 'complicit' in their own collection (Staddon 2009b). Plants can be 'message bearers for the endless chain of biophysical processes at work in the world' (Hitchings 2006: 373). Plants also offer competition and challenge to human purposes, particularly in the form of weeds (Power 2005) or invasives, when their agency is disorderly and subversive (Ginn 2008). Their difference and mobility engenders a certain sort of 'awkwardness' in the way that people talk about them (Hitchings 2007). Nevertheless we consider that these studies have left unexamined the question of what it is to be a plant, and this is something that now needs to be more explicitly interrogated.

We ourselves have previously argued that animal geographies have attracted greater scholarly effort because they have been spurred on by questions of ethics (Head and Atchison 2009). Between plants and humans, there is a greater ethical distance, we argued, and the unit of ethical standing is more in question. Our assumption of ethical distance needs to be challenged, because the systematic analysis of differences between plants and people leads us to relations of bodily intimacy as well as distance. The unit of plant subjectivity and ethical standing is an important question we return to at several points in the book. Individual plants have received considerable scholarly attention when they take the form of trees, wielding symbolic power and claiming some degree of ethical standing through size, age or importance in the landscape (Rival 1998, Jones and Cloke 2002, Cloke and Pawson 2008). In this form, plants may have become 'subjects and stakeholders' as Wolch (2007) argues for animals, but this is an exception. More often we think of plants not as individuals but as different types of collectives or assemblages: forests, biodiversity, vegetation communities. As Hitchings and Jones (2004: 5) argued, 'there is something... about the biological properties of plants that makes for an uneasy mixture of collective landscape and independently struggling organism'.

In our recent review (Head and Atchison 2009), we identified *food* as one of the collectives that had generated considerable attention in what could be thought of as human-plant geographies, particularly by those using a methodological strategy of 'following' the plant (Cook et al. 2006). In the best of these inquiries plants emerge as a 'lively presence ... agglomerating diverse acts ... complicating the distribution of knowledge and power, enabling us to "flesh out" the spaces in-between the process of food production and consumption' (Whatmore 2002: 142).

In the same way that chickens can be understood as bundles of social relations (Watts 2005), plants can also link people and places where citizens are 'unknowingly connected ... (and) entangled' (Cook 2004: 662). In most of these studies the ethnography is necessarily multisited; the researcher, in Cook's terms, 'follows the thing' (Cook 2004, Cook et al. 2006). As a number of writers have noted, methodologies that bring the consumer better knowledge of the producer make it harder to avoid questions of ethics. But the question posed here is not usually about the ethical responsibility of the consumer to the plant itself, rather to the social and ecological conditions of its production.

A second collective we identified is *garden*, a site where recent research documents a rich variety of people's everyday embodied interactions with plants. It is particularly in the domestic garden that a certain sort of plant charisma draws in human attention and care, as documented in the various works of Hitchings cited above. Thus beautiful flowers bring aesthetic pleasure and fragrance and draw people to look at and care for them. Particular plants provoke memories of the person who gave them as a gift or in whose memory they are planted. Lawns are enjoyed for the cool feel on bare feet, or hated for the ecological pathology that they are seen to represent. Trees are highly valued for shade and beauty, but can also be a source of neighbourly conflict as roots excavate sewer pipes and leaves make a mess over the fence. Vegetables provide a source of food and a connection to notions of life and productivity (Head and Muir 2007).

Recent research on people and plants intersects in productive ways with wider debates in cultural geography about *migration*, *identity* and *belonging*. Indeed we suggested they could have fitted under Blunt's (2007) encapsulation of 'mobility, transnationality and diaspora' in migration studies. Kull and Rangan (2008) recall the history of colonialism as traced in older plant geographies (Crosby 1972, 1986). By tracing the movement of Acacias between Australia and other regions of the world, Kull and Rangan illustrate how plant stories can complicate any simplistic unidirectional reading of colonial histories. Human geographical and anthropological work on invasives, aliens and weeds has drawn attention to the cultural contingency of these supposedly straightforward ecological categories, and their variable relationships to questions of national identity and exclusion of Others (for example, Ginn 2008). Barker (2008) shows how plants can disrupt human boundaries around the native and non-native by bringing other plants into their care and protection. We will see in several later chapters that wheat is not immune from these questions of belonging.

Biogeography: Finding a Place for the Human

Within physical biogeography, plants are usually approached as collectives, the bigger assemblages that make up the landscape or the habitat for humans and other animals, and that can be mapped using an increasingly sophisticated range of techniques. So the scale of analysis is an important variable in how human-plant relations are approached within biogeography. But if, in the humanities, the plants have been backgrounded as the passive backdrop of human endeavour and domination, the hierarchical ordering of the world emerges in a different way in biogeography. In a subdiscipline that takes as its subject matter the biota – usually understood as plants and animals – the problem has been what to do with the human. We take two examples, the concept of human impacts, and the recent revision of global biomes to include human activities.

Our increased understanding of *Man's Role in Changing the Face of the Earth* (Thomas 1956) is one of the key scientific achievements of the second half of the twentieth century. Geographers contributed to this achievement by showing diverse examples of the spatial and temporal reach of human activities, both far back into prehistoric time and into all corners of earth surface processes. This is seen most dramatically in the example of anthropogenic climate change.

The hard won concept of human impacts is now empirically incontrovertible and is gaining increasing political traction. Yet it continues to position humans as being 'outside' the system under analysis, as outside nature. The dichotomy between the social and the natural is *ontologically incomplete*. By positing two different spheres of reality, it leads to a conception that entities are 'essentially' either social *or* natural prior to their interaction with one another, what Latour (1993) describes as the 'modern' worldview. Even the concept of human 'interaction' with environment – a milder version of human impacts – is also problematic in this understanding, since it retains the assumption that the social and the natural are pre-existing categories prior to their interaction with one another.

And yet, in biogeography there are many good empirical examples of studies that are spatially and temporally finegrained, that deal with complexity and contingency, and that likely acknowledge multiple agency. A number of these are at least partly consistent with 'a conception of action and actors which is multiple, contingent and nonessentialist' (Castree 2002: 121). There are several elements of common ground here. This understanding of power and agency has much in common with recent approaches in ecology, in which change and contingency rather than stability is the norm, and 'disturbances' such as fire and human actions are understood as internal to the system rather than external (Hobbs et al. 2006, 2009, Davis et al. 2011).

The encouragement to think and explain relationally, in terms of associations rather than separations, is a subtle but profound challenge. A relational perspective requires us to ask much more about and elaborate in detail the specific mechanisms of connection that prevail in particular times and places, both past and present.

Empirical evidence, reflecting plants in vigorous encounter with humans, has caused biogeography to revise its own previous categorisations. Traditional biome definitions ignore or exclude humans and crops; for example 'a *biome* is a large ecosystem in which relatively uniform climatic conditions lead to uniformity in the living-organism environment, particularly in the type, or growth form, of the primary producers' (Bradshaw and Weaver 1993: 572). A recent revision of global biomes is arguably the most important and systematic acknowledgement, in the last few decades, of human agency in vegetation systems, or global ecosystems (Ellis and Ramankutty 2008). Ellis and Ramankutty (2008: 440) characterise 18 'anthropogenic biomes', based on an empirical analysis of population, land use and land cover at 5 arc minute resolution (~86 km^2 at the equator), 'a spatial resolution selected as the finest allowing direct use of high-quality land-use area estimates'. Examples include 'rice villages', 'residential irrigated cropland' and 'populated forests' (Ellis and Ramankutty 2008: Fig. 1). Although it remains a conceptual model, croplands are characterised (their Fig. 3b) as having high net reduction of biodiversity, a few introduced species and minimal surviving native biodiversity.

Their aim in doing this revision is to set up testable hypotheses to facilitate more accurate models of global change, for example by including field measurements of net primary production and carbon emissions. Partly this is a scale issue. 'Observations within anthropogenic landscapes capable of resolving individually managed land-use features and built structures are critical, because this is the scale at which humans interact directly with ecosystems and is also the optimal scale for precise measurements of ecosystem parameters and their controls' (Ellis and Ramankutty 2008: 445).

What do these new maps say about human-plant relations? They represent a significant shift in biogeography in terms of bringing humans into conceptual schema which hitherto excluded them, even if the terminology still reflects humans being outside and separate from something called an ecosystem. And humans are implicitly positioned only as agents of damage, rather than also of enhancing biodiversity (Rival 2006). Nevertheless, this is the most systematic attempt to render visible the extent to which human presence and processes have become embedded in the structure of biomes. It has been driven by the bottom up empirical evidence of a transformed earth (although of course there is still much to find in the human-plant relationship below the smallest pixel size of 86km^2).

There are many other examples of biogeography doing well with the empirical evidence of a human-infused world because it proceeds from bottom up rather than imposed top down categories (see Sagoff 2003 for more on these as different concepts of ecological science). Examples include Niggemann et al.'s (2009) consideration of the important human role in seed dispersal via cars, clothes and shoes; the characterisation of vegetation heterogeneity in urban ecosystems, illustrating that 'vegetation is one of the three main spatial structuring elements of urban areas' (the other two are buildings and surfaces) (Pickett et al. 2011: 355); Hobbs et al.'s (2006, 2009) conceptualisation of novel ecosystems; Ladle and Jepson's biocultural theory of extinction (2008); Atchison's (2009) analysis of fruit trees in monsoon savannahs and Laris's (2011) integration of human practices into ecological models of savannah fires.

Agriculture and Domestication: Thresholds and Critique

The evolution of agriculture, sometimes referred to as The Neolithic Revolution, is consistently understood as a threshold moment in human history. It significantly increased the availability of calories per unit of land and labour invested. The storage and trade of significant food surplus paved the way for a transition from hunter-gatherer society to sedentism, in turn providing the necessary population growth for cities and the emergence of civilisation. Agriculture led to widespread transformation of the face of the earth through the processes of land clearing. More recently, it is understood as having transformed the carbon budgets of human ecologies. For example Ruddiman (2003) has argued that early agriculture marked the beginnings of human influence on the atmosphere through methane release from rice irrigation.

This story is often told in a linear and determinist way that seems to emphasise the inevitability and superiority of agriculture sweeping across human history. However, increasing evidence from the archaeological record documents enough spatial and temporal variability in this process to challenge the coherence of the cultural and economic package we call 'agriculture'. Indeed 'it is unlikely to have hung together as a concept without the central notion of separating humans/ culture/civilisation out from nature' (Saltzman et al. 2011: 56).

Agriculture also differentiated those peoples who had culture from those – hunter-gatherers, primitives and savages – who did not. The concept and practice of agriculture can be understood as central to the emergence and maintenance of the culture/nature dichotomy within Western thought and practice. Knobloch (1996: 74–75) traces the etymology, arguing that '"culture" appeared during the great reclamation of the sixteenth century and meant "agriculture", although in a few years "agriculture" was a word of its own'. The companion concept of domestication, usually marked by morphological changes in animals and plants as a consequence of human intervention in breeding cycles, also contains this conceptual shift from nature into culture. Domestication, particularly of animals, is understood as taming the wild and bringing it into the human sphere (Anderson 1997).

In this section we briefly review three interwoven aspects of the emergent critique of agriculture and domestication – *conceptual, empirical* and *material* – focusing on evidence which has implications for human-plant geographies. Expanding ethnographic and ethnohistoric research into hunter-gatherer lifeways in the second half of the twentieth century revealed many examples of practices previously associated only with agriculture, gardens and cultivation. Examples include the encouragement of fruit seed germination on the edge of Aboriginal campsites (Jones 1975: 24), both extensive and small scale sites of yam cultivation (Hallam 1989, Lucas and Russell-Smith 1993), and many descriptions of tilling the soil to enhance the flourishing of tuberous food sources (Gott 1982). Chase (1989: 48) called these practices domiculture, the 'knowledge and activity bundles which relate temporally to a specific habitat'. Another series of influential papers examined subsistence strategies across the boundary zone of Torres Strait, using it as a transect between the hunter-gatherer groups of northern Australia and the agriculturalists of New Guinea (Harris 1977). Harris expressed this spatial variability as a continuum of human-plant relations (Harris 1989, 2007), whereby domesticated species could be important to a greater or lesser extent dependant on the relative significance of that food or species.

Empirical evidence of hunter-gatherer cultivation processes was often ignored or rendered invisible in the complex process of colonisation. This was partly to do with their gendered nature; the descriptions are overwhelmingly of women's work (Gott 1982, 1983). Both Knobloch (1996) and Anderson (1997) point to the ways these were raced and gendered ideas from the beginning. In a number of New World contexts, the agricultural metaphor was central to the colonising culture's vision of itself and its civilising presence. 'Improvement' of the land was related to the transforming hand of civilised man in the form of land clearing, followed by the plough, the herd and the fence. A process of conceptual dispossession attended the physical dispossession (Head 2000, Anderson 2003).

Conceptual critiques of the hunter-gatherer agricultural dichotomy came from anthropology, with Ingold's (2000) articulation of dwelling, and from geography with Anderson's (1997) critique of animal domestication. Building on examples of how 'others' conceive of their relationship with plants, Ingold reconceptualised human-nonhuman relations as being the 'relative scope of human involvement in establishing the conditions for growth' (Ingold 2000: 86), without making distinctions between the natural and social worlds. Anderson synthesised an 'appeal' to relax rigid oppositions and reframe 'and re-imagine more animal-inclusive models of social relations' (Anderson 1997: 463). She argued that the 'underpinning moralities and contradictory manifest forms' of domestication are open to 'rupture and reversal' (481). Scholars across a range of disciplines have taken up this challenge, producing new accounts of human-animal relations in which the boundaries previously drawn are not so distinct, and in which the human cannot be privileged in quite the same way (see for example Cassidy and Mullin 2007). Within archaeology, reconceptualisations of domestication as a social and cultural process, rather than just a rearrangement of genes (Denham 2007a, 2007b, Hodder 2007) are other examples of this trend.

Also within archaeology, there has been a wider rethink of the Neolithic Revolution, and its Near East centre of origin, over the last two decades (Thomas 1991). Evidence increasingly showed that the various parts of the Neolithic 'package' did not all occur together, nor necessarily always in the same order. Even within a region such as Europe, sedentism sometimes preceded, sometimes followed agriculture. Further, agriculture had emerged independently, in different configurations, in different parts of the world, including New Guinea and the Americas. Evidence from yams, taros and bananas, for example (Denham 2007a, 2007b, Vrydaghs and Denham 2007), challenged the dominant 'cereal-centric' models.

Jones and Brown (2007) show in detail how the morphological changes to plants and animals, and the set of practices documented from the Near East, have come to dominate thinking about the origins of agriculture, arguing that that area has defined the tests for both empirical evidence and the frameworks for thinking about subsistence and food production. For example, the specific morphological changes seen in domesticated plants, particularly gigantism and dehiscence (the spontaneous opening at maturity of a plant structure, such as a fruit, anther, or sporangium, to release its contents) in wheat and other cereals, characterise expectations of how domesticated plants could be visibly (and genetically) distinct and different to their wild counterparts. The focus on Eurasian cereal agriculture, which includes the story of the domestication of wheat, is argued to fetishise the significance of morphological changes at the risk of ignoring or underplaying more significant social and ecological change (Denham 2007a, Denham and White 2007, Vrydaghs and Denham 2007). In this view, morphological change is an 'artificial' moment in time – a point only along the line of an evolving relationship between the humans and plants. Some relationships might be quick and dramatic; others slow and evolving; some intense or indeed with little commitment from either human or plant partner (Zeder 2006).

This empirically driven critique draws on a more diverse range of evidence than had been available before. Archaeologists by definition work from material evidence, so they naturally focus on morphological changes in plants, such as the change in sizes of fruits or grains. However Vrydaghs and Denham (2007) argue that some agricultural practices do not yield an 'anticipated' domestication signal. A wider range of approaches is necessary, both to look at plant remains in different ways, and to examine other parts of the agricultural 'package', such as associated social changes. New techniques such as molecular and residue analyses are providing evidence of spatial, temporal and functional diversity: 'there were mosaics of different plant and animal exploitation practices juxtaposed across space, which were variously transformed through time. Plants, technological and other elements of material culture dispersed along local exchange pathways as well as with movements of people' (Vrydaghs and Denham 2007: 5).

The range of new evidence and perspectives does not mean that archaeologists have abandoned the concept of domestication, or its significance as a threshold, completely. For example Harris continues to argue for, and defend the place of, domestication as a defining characteristic of agriculture because 'it represents a fundamental change in human subsistence [which] eventually led to increased density of human populations per unit area' (Harris 2007: 20). Others have attempted to do away with the term domestication (Terrell et al. 2003), redefine it (Zeder 2006) or use it in a more fluid sense (Cassidy and Mullin 2007).

We find it helpful to consider human-plant relations over archaeological timescales as constituted by 'bundles of practices', and to be reminded that close empirical attention to variation in space and time reveals very different patterns to the imposition of categories (Denham et al. 2009). It is also important to consider where and when wheat may have been overplayed in the historical story. The challenge for us is to how to decentre wheat when necessary, even while centring it as the core narrative of the book.

What is a Plant?

For all their excitement and liveliness, these productive recent discussions about human-plant relations have sidestepped the fundamental question, 'what is a plant?' With colleague Catherine Phillips we have considered this question and its implications (Head et al. 2011, Phillips in prep.). We were struck by the difficulty of going further in understanding the particular qualities of planty relating until the question was systematically addressed. (It is of course also necessary to ask, what is a human? But that question has received much more scholarly attention, underpinning the humanities themselves.) We approach it here in two very different – arguably opposing – contexts, Western science and Indigenous worldviews. On the face of things scientific taxonomy proceeds on the basis of separation and differentiation, and the Indigenous perspectives we draw on proceed by association and relationality. In the latter context the question 'what is a plant?' may not even make sense. However, there are also grounds of connection between the two views if we think of them as ways in which people have engaged with, and tried to make sense of, plant materiality and capacities.

The biological answer to the question 'what is a plant?' is not straightforward, since even undergraduate texts readily acknowledge much more ambiguity and difficulty than Aristotle apparently encountered (Simpson 2006: 3). One problem is that the question risks implying an essence based on individual and particular forms; what counts as a plant is contingent and has evolved over time. Nor is it the same question as *what do we recognise as a plant?* The material expressions of plants with which we are most familiar – leaves, seeds, bark, roots, stems, flowers – are relatively novel in evolutionary terms.

To frame the question in this way is already to step into the assumptions of western science, and to provide answers based on scientific taxonomy. Plants are defined as living organisms belonging to the Domain Eukarya,[1] Kingdom Plantae. So it is important to recognise the nature of taxonomic practice in the construction of categories in and around plants. Any system of classification is a process of stabilising the representation of entities, simultaneously emphasising both difference and sameness (Geissler and Prince 2009). Scientific taxonomy orders things based on a combination of their physical characteristics and their evolutionary relationships. Taxonomic levels are themselves relational fields – they do not pre-exist their members. The taxonomist, working from the materiality of the plant, brings a new taxonomic category into being when necessary. But as Margulis and Schwartz (2009: xiv) remind us, taxonomy is a constantly changing and evolving process. As new cellular, sub-cellular and genetic information is added, older categories and relationships often need reinterpretation.

The question *what is a plant?* soon became for us a more useful, albeit clumsy, question: *what is plantiness?* We advance the concept of *plantiness* to distinguish

---

1   Eukaryotic organisms are distinct from Procaryotic organisms in that their cells have nuclei, and membranes surround their subcellular structures (organelles).

between the capacities that plants have in common, though they manifest in different forms, and those, such as flowers or leaves, which are confined to particular groups of plants. Plantiness is an assemblage of the shared differences of plants from other beings. No single characteristic can be used to capture and define plantiness. Rather it is usually an assemblage of the following characteristics:

- undertaking photosynthesis;
- being multicellular;
- having predominantly cellulose walls;
- storing energy as starch; and,
- having an alternation of generations in their lifecycles (sporic meiosis).

The process of the emergence of what we call *plantiness*, as understood in evolutionary history, is summarised as follows. Unlike animals (known as heterotrophs) and fungi (saprotrophs) which must obtain their energy from other organisms or the surrounding environment, and some forms of bacteria, which can obtain energy from surrounding chemicals in their environment (chemotrophs), some organisms capture the potential energy of light and store it within their bodies (phototrophs). Phototrophs carry out the chemically complex and variable process known as photosynthesis, where incoming light energy from the sun is captured and stored as chemical energy. As the dominant form of energy capture and storage, photosynthesis underpins feeding and food (trophic) relations among the vast majority of living things.

The ancestors of modern photosynthesising bacteria (cyanobacteria, sometimes known as blue-green algae) were probably the first living organisms to undertake photosynthesis using water as the electron donor (instead of sulphur or other elements which are used by other bacteria). This was an immensely significant evolutionary step, which initiated the release of oxygen originally bound up in the earth's water and eventually resulted in the oxygenation of the atmosphere, making larger and more complex organisms possible (Graham et al. 1995). Today a suite of organisms, including some bacteria, algae and plants, carry out photosynthesis.

Using a range of molecular and genetic evidence, some scientists argue that unicellular eukaryotic organisms incorporated ancestral cyanobacterial cells into their bodies between 1–0.8 billion years ago (bya), acquiring via endosymbiosis the capacity for photosynthesis (Ennos and Sheffield 2000, Raven and Allen 2003). This origin is recognised today in chloroplasts (green[2] plastids) – the subcellular units where photosynthesis takes place in both green algae and in plants. At a molecular level, sunlight splits water molecules inside the cellular structure of the

---

2 The chloroplasts in green algae and plants are green because they contain both pigments chlorophyll a and b. Chlorophyll a is common to all photosynthetic organisms; while only green algae and plants possess chlorophyll b in addition. Other algae also possess other pigments which are possibly an adaption to low light conditions of deeper marine environments.

plant, releasing oxygen, and then cleaving hydrogen to carbon dioxide molecules, thus assimilating and binding (chemically reducing) atmospheric carbon into sugar (glucose). Photosynthesis[3] also makes possible the production of more complex sugars and other compounds, notably starch (the insoluble complex carbohydrate which can be accumulated and stored), and cellulose (a polysaccharide chemically similar to starch, but with slight differences in the arrangement of chemical bonds). Cellulose becomes fibrous, giving structure, strength and form to plants, and is the most abundant biopolymer on earth (Cosgrove 2005: 850).[4]

Between about 0.8 bya and 500 million years ago (mya), multiple evolutionary 'experiments' took place. Some unicellular organisms (primitive algae) persisted with free floating planktonic habits; some began clumping together to form multicellular units; and some organisms took to benthic or bottom dwelling life (Ennos and Sheffield 2000). Multicellularity developed independently in divergent organisms a number of times, and may have developed in some algaes as early as 1 bya (Xiao et al. 1998), but Ennos and Sheffield (2000) argue that modern terrestrial plants were most probably derived from very simple multicellular planktonic green algae (Chlorophyta). Contemporary biologists recognise a huge diversity in the form and functional types of algae, from simple single celled organisms to multicellular highly complex organisms such as kelp, seaweeds and coralline algae; but there is continuing debate and disagreement about how to classify and describe them (Cavalier-Smith 1998, Lewis and McCourt 2004, Simpson 2006). Although the evolutionary relations and functional similarities to plants are recognised, the strictest systems of systematic taxonomy continue to classify all algae in the Kingdom Proterozoa (Protista), not Plantae.

Some time before the period 450–470 mya, particular algae – most likely from the group Charophyceae (Graham 1996) – made the transition from purely aquatic to terrestrial life. This transition was made possible via the evolution of a novel way of ensuring safe dispersal and delivery of reproductive cells (gametes), involving first the alternation of generations (also known as sporic meiosis), and second the production of sporopollenin. Sporic meiosis is the most complex form of multicellular reproduction and is specific to Charophyceae algae and Plantae. These organisms produce two distinct multicellular organisms at different stages; the diploid organism (most commonly a sporophyte[5]) and the haploid organism

---

3   Photosynthesis is often represented by the following equation but even this simplifies a more complex set of cyclical and stepped processes (see for example Rost et al. 2006). $6CO_2 + 6H_2O > (light) C_6H_{12}O_6 + 6O_2$.

4   Simpson (2006) notes that the production of starch represents a unique feature of both green algae and plants but that there is some uncertainty about the evolution of cellulose production which may have evolved previously and thus may not constitute a unique feature (apomorphy) for green algae and plants.

5   The spore (or sporopollenin) producing stage of the life-cycle. See Chapter 3 for further explanation of ploidy in relation to wheat.

(most commonly a gametophyte[6]) (Bennici 2008). In Charophytic algae, these life stages are equidominant (Ennos and Sheffield 2000); in plants one of these two stages has become the dominant physical life form, with the alternative form diminutive and mostly dependent on the larger. In mosses and liverworts for example, the gametophyte is the dominant life cycle; in contrast the sporophyte is the phase most of us would recognise as the dominant life form of a flowering plant (Ennos and Sheffield 2000).

Sporopollenin is a tough resistant coating around the minute capsules (spore and pollen) containing the gametes. While Charophytes had been dependent on aquatic media to perform this exchange, terrestrial plants had to contend with periodic or total desiccation. The development of protective sporopollenin thus enabled the spores and pollen of terrestrial plants to be widely dispersed away from the immobile parent organism; allowing future generations to overcome potentially restrictive or difficult local conditions (Kinlan and Gaines 2003).

Thus we can understand plantiness as having come together in various forms by around 470 million years ago. It was quite some time later (approx 360 mya) before specialised sexual organs, specialised vascular, structural and growth tissues, stomata, leaves, roots and upright tree habits emerged (Henrick and Crane 1997). Between 280–360 mya, extensive forests of the earliest seed plants dominated the land and were laid down in coal seams. By 90 mya flowers developed, and the plants with which we are most familiar came to dominate plant life on earth (Qui et al. 1999).

To state the obvious, plantiness long predates humans. In fact, whatever humanness is, it requires plantiness. We are made by plants in the sense that they have provided the atmosphere that we breathe and provide much of the sustenance that we eat. They have had agency in the ways our bodies evolved, and continue to be fundamental to our daily bodily relations. In making this argument, however, we are mindful of Kearnes's critique that the material can operate as 'a sign for *the natural*, the pre-discursive or the *a priori*' (2003: 144). Kearnes argues instead for the expressive potential of matter, understood as 'simultaneously – and unevenly – discursive and physical' (Kearnes 2003: 150). This can be seen in the way our understanding of plantiness – as an assemblage that combines material characteristics with processes and characteristics – has itself been derived via the modes of representation of scientific thought.

The specifics of plantiness help us to consider the specifics of plant agency – how it comes out of certain material capacities – and how that prefigures relations with people. To take just one example, many different plants lay down large stores of starch and, in response to the marked seasonality of the monsoonal tropics, plants with tubers store starch very efficiently. Coursey (1973) showed that hominids knew about and took advantage of tubers as a rich source of carbohydrate (a trophic, metabolic relation) very early on. Archaeological and anthropological research over the past thirty years suggests, further, that starch from tubers played

---

6   The gamete producing stage of the life-cycle.

a crucial role in the particular course of human evolution and development. Wrangham et al. (1999) argued that the innovation of cooking starch resulted in the evolution of a unique human social system of pair bonding. Similarly O'Connell (2006) articulates how starchy tubers may have enabled early weaning of children and in doing so increased fecundity and promoted delayed maturity of human populations. We return to archaeological evidence below, to discuss a further important threshold in human history, the evolution of agriculture.

To fully consider the plantiness of plants, and plant engagements, it is necessary to rethink concepts such as agency, mobility and intelligence more fully than is possible here (but see Hall 2011 and Phillips in prep. for more detailed discussion). We briefly discuss here the debates around plant intelligence for their connection to the question of individuals. Trewavas (2002, 2005) has argued specifically for plant intelligence to be recognised through plants' many reproductive, adaptive, communicative, planning and predictive capacities. The internal workings of plants are communicative; through assemblages of (among other things) proteins, minerals, and chemicals, carrying complex signals to various cells and tissues (Trewavas 2002) in which learning and memory develop (Trewavas 2005). The Venus fly trap, for example, can be said to have sensory memory similar to animals in its ability to sense, react to and trap its prey. The rapid closing of the leaves (or trap) occurs when at least two sensor hairs respond to stimulus and chemicals are released, signalling the leaves to close (Ueda et al. 2007, 2010). Debates about chemical signalling challenge our ideas about passivity, by suggesting that plants perceive, process and react to environmental information.

Not surprisingly, this has been a controversial debate, partly because of the close association of intelligence with human-centred concepts such as the mind. More specifically, Firn (2004) argues that 'as intelligence is a property of individuals, plants cannot be intelligent as they are not individuals in the same way as animals' (Hall 2011: 145). Hall agrees instead with Trewavas, arguing strongly that plants are individuals. This provides the basis for Hall's later argument that it is these 'individual plant persons' (p. 169) with whom humans should work collaboratively. Because this concept of the individual is one that we find insufficient in considering the full gamut of human-plant relations (particularly where wheat is concerned), it is important to consider exactly what is meant by plant personhood, as articulated primarily within indigenous traditions.

Indigenous Plant Understandings

For millennia people have had diverse ways of knowing and understanding, ordering and relating to living organisms, including some in which the boundaries between humans, animals and plants are not drawn as clearly as in the West (for example Terashima 2001, Mosko 2009). These understandings have three particular areas of relevance to the present study. The first is a methodological one, the consideration of what 'ethnobotanical' research might look like, and how it might be done, in the context of contemporary wheat. Indeed this project sprang indirectly from our own previous research into Aboriginal plant use, including yams and wet season fruits, in monsoonal north western Australia (Head et al. 2002, Atchison et al. 2005). Working over a number of years with Aboriginal women on country challenged us to more systematically examine the environmental relationships, including those with plants, we take for granted in our own mostly urbanised lives.

Second, it is necessary to acknowledge the Western cultural specificity of our particular concern with human-plant commonalities and differences. As Strathern (1996: 525) has argued, the idea 'that social relationships concern commonalities of identity before they concern difference, and that heterogeneity is inevitable in combining the human with the nonhuman' is a particularly Euro-American cultural predisposition. Strathern contrasts this with networks, such as among the 'Are'are of the Solomon Islands, 'that are homogeneous in so far as they presuppose a continuity of identities between human and nonhuman forms, and heterogeneous in so far as persons are distinguished from one another by their social relationships' (Strathern 1996: 525). More recent anthropological writings discuss the concept of personhood as it applies to other-than-human things, including plants. Hall reviews some of this 'new animism' literature, quoting Graham Harvey's definition of persons as 'those with whom other persons interact with varying degrees of reciprocity. Persons may be spoken with. Objects by contrast, are usually spoken about. Persons are volitional, relational, cultural and social beings. They demonstrate agency and autonomy with varying degrees of autonomy and freedom' (Harvey 2005: xvii, quoted in Hall 2011: 105).

In the Australian indigenous context, this sentience and subjectivity is accorded to country, as elaborated by Rose:

> Country is a place that gives and receives life. Not just imagined or represented, it is lived in and lived with. Country in Aboriginal English is not only a common noun but also a proper noun. People talk about country in the same way that they would talk about a person ... [It] is multi-dimensional – it consists of people, animals, plants, Dreamings; underground, earth, soils, minerals and waters, surface water, and air ... Country has origins and a future; it exists both in and through time. (Rose 1996: 7–8)

The sentience of country can survive the surface transformation from a bush context to a garden form, as we described for Murinpatha woman Biddy Simon's outstation garden (Head et al. 2002). Plantings included edible yams and tubers collected from the bush, as well as *Nauclea orientalis* (Leichhardt pine) seedlings collected from a rainforest patch some tens of kilometres away. We argued that, in bringing bush plants back to her garden, Biddy was actively maintaining connections with different parts of country; she remembered and talked about when and where she got them. Nor was it only food plants that were grown. Bamboo and other similar reeds provided spear shafts for fishing. From her garden, Biddy observed many details of her environment, such as the number and type of birds visiting the adjacent billabong. Although this looked like an enclosure of space, a creation of 'inside', surrounded as it was by a suburban wire fence, we interpreted it as just another aspect of Biddy's close engagement with country. Plants were often referred to in relation to their smell, e.g. 'smellem from long way'. Country also talks: 'Listen all the sugarbag [bush honey] singing out', she instructed us, in reference presumably to the bees. Since country has an active presence, and is engaged with using all the senses, it must also be part of domestic space.

Rival (1993) drew attention to the way plant temporalities, expressed in three different kinds of trees, intersected with human ones among the Huaorani of the Amazon, informing understandings of growth itself. The Huaorani 'draw a fundamental distinction between living organisms that grow slowly and perdure in groups, and those that grow fast but die off' (Rival 1993: 648).

We are not arguing that indigenous understandings of plant personhood or sentient country translate in straightforward ways, if at all, to our analysis of wheat. Nor is it only in indigenous understandings that concepts of plant personhood can be found. Hitchings argued that 'the plants performed themselves into existence as discrete entities such that they became almost considered as similar to people' (2003: 107) in the context of London gardens. But the third area of relevance is that cross-cultural comparisons are an important means to understand the contingencies of how some relations become embedded and others not. They also provide inspiration to consider how things can be done differently.

Towards a Human Bio-geography of Wheat

What can we take from all this work to help us re-approach human-wheat relations? For most of us who live in urban areas this is the least care-ful of our plant relationships – we are likely to attend to the flourishing of flowers in the garden, the herbs on the window sill or the tree in a neighbourhood park with far more thought and feeling than wheat. It is an industrial, commodified relationship that apparently takes place somewhere else. Is it really relevant then to think of this first and foremost as our relationship with a plant or groups of plants? This apparent – but false – distance is precisely the reason why this mundane relationship can be fruitfully rethought in these terms. Hall leaves much unsaid on agricultural plants; his focus is rather on free and autonomous plants, for good reason: 'the conversion of autonomous plant habitats into agricultural areas intended to satisfy human purposes threatens the integrity of plant species, ecosystems, and the biosphere as a whole' (Hall 2011: 164). At this point we could invoke the heritage of Theophrastus, who understood 'cultivation to be a collaborative, mutualistic, relationship between plants and humans' (Hall 2011: 35) in which there were both detriments (for example suffering of cultivated trees as they are pruned) and benefits (plentiful food and water, removal of competitors) for the plants. However, what stands for reciprocal care and responsibility in a

Greek olive grove in the third century BCE may be very different from a twenty-first century wheat farm, and the globalised market into which its harvest flows.

This issue is further complicated by the way in which wheat has helped to constitute Western understandings of property and markets. That is to say, wheat is not just an 'example' of something that circulates in a bigger entity called global agricultural trade, it is rather ingrained in the very conceptualisation of that larger entity. Mitchell (2002: 85) attributes 'the idea that a country's social and economic relations can be pictured in terms of agrarian property' to the nineteenth-century economist David Ricardo, for whom 'the dynamic of creating wealth began not with the act of exchange, but with the process of settling and cultivating an empty land, a space of colonization.'

> … in England, when Ricardo first outlined a simplified idea of the making of wealth as the expanding control of agricultural land and the consequent increase in agricultural income, his model of the circulation of wealth was based on the cultivation and consumption of a single product, wheat. With the expansion of large landownership in England, the wheat crop had replaced a more diversified grain agriculture and played a dominant role in farming, trade, and consumption. (Mitchell 2002: 94)

'Writers like David Ricardo described a regular motion of production, exchange, and consumption whose regularity derived from the natural cycle of the country's major commodity, wheat, and whose movement they called the market' (Mitchell 2002: 246).

This view of wheat now flashes across our TV screens together with the price of gold and oil. It seems a long way from the plant. Yet if we abandon agricultural plants to the realm of culture (or politics, or economics) just at the point where scholars are finding new ways to approach human-plant relations, we reinscribe old boundaries and miss the opportunity to include agricultural staples in discussions of how to live more ethically and sustainably. In the remainder of the book we draw on and advance these theoretical perspectives in the following ways.

Plantiness

Many millions of years before it entered an intimate relationship with people, wheat's ancestors had an independent life as grasses. Hanging on to the concept of plantiness helps us attend to the materialities and capacities of plants in their own terms. This is just as important for agricultural plants as for wilderness trees. Crops have not lost their plantiness, notwithstanding that humans are intervening in and manipulating that plantiness in increasingly intrusive ways. All plants, even agricultural ones, have some plantiness that is independent of humans, or at least beyond the control of humans. Aspects of plantiness, such as starch or cellulose, will be more or less important but nevertheless present.

Becoming

Considering plants on their own terms does not deny the importance of human activities. Humans intervene in different parts of the life cycle in crops and garden plants, often in ways that render the plant dependent on continuing human engagement. Several temporalities intersect here. Rival notes that domestication 'presupposes dependence on plants whose growth is much faster relative to human growth and maturation processes' (1993: 648) than the long-lived forest trees of the Huaorani. The insights of archaeology, palaeoecology and evolutionary biology help us understand how wheat enfolds many thousands of years of past relations into contemporary lives. This becoming is a continual process – just as contingency and intentionality have interacted in the past, so they can in the future. Becoming is a concept we apply to plantiness itself; we try to account for 'the stability of current actors not through reference to some eternal 'essence', but historically layered contingencies' (Ginn 2008: 8).

Methods in the Ethnographic Tradition

In studying human-plant relations, particular combinations of agency, categorisation and practice are not known in advance but are to be approached empirically (Hitchings and Jones 2004). Consistent attention to practice rather than theory allows an understanding of 'multiple matterings' to emerge; matter is being 'done' in many different ways (Mol 2002, Law 2010). The construction of scientific taxonomies, the gathering practices of indigenous women, and the suburban gardening experience can all be understood as practice-based engagements with the materiality of plantiness. Methodologically, we followed both wheat, and wheat people. Our human cohort involved some 80 participants, including 29 farming families.[7] How might we count the wheat? In following wheat, we were attuned to its rhythms and temporalities, such as the seasonal cycle of growth, abundance and harvest (or in the case of our main field seasons planting, lack of growth and death). But we were wary of trying to speak for wheat, and cautious that ethnography is an irretrievably and proudly human-centred method. For all our angst, there is no alternative to starting in the middle of things. As Roe (2006a: 109) says, 'You cannot imagine talking to a carrot, but you can imagine a carrot in a larger network of fields, farms, industrial processing and supermarket-shelving.' As things turned out, hearing people talk about wheat and watching their embodied engagements with it gave a certain sort of voice to the wheat. Wheat inserts itself into these conversations, but it does so in many forms – as the stretch in the dough, the dust in the stock feed mill, even the soaring price on the Chicago Board of Trade. Edensor (2010: 7) argues, 'a human, whether stationary or travelling, is one element in a seething space pulsing with intersecting trajectories and temporalities', many of which are nonhuman. Those intersections

---

7   Pseudonyms for people, places and organisations are used throughout the book.

provided our ethnographic windows, the diverse space-times of encounter around which the book's narrative is structured.

Identity and Collectivity

Following wheat has built in safeguards against any essentialist conceptualisation of its identity, because it changes form before our eyes, and in hidden ways. It is necessary then to acknowledge and problematise the ways we collectivise nonhumans – the diversity of 'individuations and groupings' (Bear and Eden 2011). As we saw above, much of the debate around plant intelligence, and plant personhood, focuses on the question of whether or not plants are individuals. For us, the unit of concern with plants is very much an open question, and we do not offer a definitive solution. As Hitchings (2003: 107) argued, 'if we think relationally, plants are not necessarily always plants. They have to work to be considered as an actual entity rather than rest on the laurels of actual physical properties'. He was talking about garden plants, which have both similarities and differences to wheat. In the case of wheat, humans engage with many different collectivities beyond the individual – food, crop, seed, commodity, industrial substance, landscape. Each of these collectivities has different combinations of entanglement, posing different but just as important ethical questions.

Webs of Relationality

To focus on the human-plant relationship is not to ignore the wider webs in which both are embedded. The people-wheat nexus includes many other companions (insects, soil, rainfall, herbicides, frost, trains, mills, ovens) with their many other possibilities of agency. These webs are also spatially extensive – we should not focus only on the farm but also on the transport, trading and processing edifices that co-constitute industrialised agriculture. But we want to start with the plant itself, and it is to this we turn in the next chapter.

Chapter 3
Becoming Wheat

Selfing

Because of wheat's centrality as a staple crop throughout the west, and increasingly beyond, its story has conventionally been told in linear and progressivist narratives that emphasise the sweep of human history, the rise of civilisation and the importance of wheat in nation-building. Although the structure of this chapter echoes a linear trajectory, we are more interested in emphasising that wheat's history was, and continues to be, a messy and contingent process, an experiment that is still happening. To emphasise that wheat is in a constant process of becoming, we draw attention to different ways in which its identity has been stabilised and disrupted.

What is wheat? This apparently straightforward plant defies our efforts at defining and describing it in several important ways. 'Wheat' is an ethnobotanical construct, a human-defined collective of plants encompassing taxonomic, genetic and morphological diversity. It belongs to the grass family (*Poaceae*, tribe *Triticeae*), but its taxonomy is enormously complicated, with all attempts at systematic classification proving very difficult (Morrison 2001, Nesbitt 2001). Nearly all modern wheat cultivars belong to either *Triticum aestivum* (bread wheat) or *Triticum turgidum* (hard durum type wheat) (Zohary and Hopf 2000). However, wheat encompasses a much larger botanical suite, including two genera, *Triticum* and *Aegilops* (the 'wheat group'), (Zohary and Hopf 2000, Morrison 2001) and approximately 600 species (Cornell and Hoveling 1998). The genetics of all ancient and modern wheats further complicate this picture as wheat comprises a polyploidy[1] series of five types: diploid (two types); tetraploid (two types); and hexaploid (one type) (Zohary and Hopf 2000: 28).

And how did it become itself? We know it now as one of the cereals – edible grasses – that sustained the rise of western civilisation. Its character and identity have evolved in company with humans and their efforts – conscious and subconscious selection, transplantation over short and long distances, exchanges with other humans, and deliberate breeding over thousands of years (Feldman 2001). The result is a plant group with such enormous genetic diversity it is estimated to be about six times larger than the maize genome (Huttner and Debrand 2001).

---

1   A haploid cell contains one set of each chromosome in each nucleus. A diploid cell contains two copies (one from each parent) of each chromosome in the nucleus. Polyploid organisms have multiple copies of each chromosome in their nuclei. Polyploidy is more common in flowering plants; it gives rise to a greater store of genetic variability and thus confers greater evolutionary potential (Toothill 1984).

To speak of wheat today is to speak of an impossibly complicated geopolitical and planty assemblage. We do not go far into trying to make sense of that complexity without meeting the temptation to cede all explanatory power to something called 'the global economy' or 'multinational trading interests' in the connection between molecular biology and international capital in breeding programs, or the conflation of food aid aspirations and flagrant national self interest in the agricultural policies of rich countries. In cutting one particular path through this complexity we have tried to do two things to be faithful to our overall purpose of keeping the wheat front and centre while also focusing on its relationship with people.

One, we consider how the plant had agency in all these processes through various expressions of its plantiness, particularly its *selfing*. In common with some other grass species, all wheats have a single botanical characteristic which significantly differentiates them from the majority of other flowering plants: they are fully self-pollinated (selfing) as opposed to cross pollinated. This quaint botanical term, selfing, bestows identity, even an independent personality, on the wheat. Selfers, or self-pollinators, are especially suitable for domestication for two primary reasons. First, selfing effectively separates domestic or selected plants from wild progenitors. This enables human cultivators to select particular seeds from plants deemed desirable and keep the crop separate, if not geographically distant, from surrounding wild populations. Second, selfing also enables the cultivator to select and keep varieties (genetically independent homozygous lines) separate from each other in the gene pool of the crop, thereby preserving varietal identity (Zohary and Hopf 2000).

Two, we try to keep front and centre the idea that all these global edifices of intimidating complexity are constituted by everyday practices. In many cases our ethnographic perspective allows us to simply highlight those everyday practices, whether it is a PowerPoint presentation at an international conference or an incidental grinding practice by a woman twenty thousand years ago.

In this chapter we capture several long moments of becoming that we consider most important in understanding the contemporary wheat context. The becoming is variously and simultaneously evolutionary, definitional, spatial, political and technological. It includes processes and categories imposed on wheat by humans, and circumstances where the plantiness of wheat dominates. We start by examining the process of *becoming grass*, exploring how the materiality of these plants evolved. This history cannot be understood as separate from scientific practice, and some of the definitional debates that emerged in the biological encounter. Then we outline wheat's *domestication*. By all accounts, wheat has one of the most complicated domestication histories being written – a story of multiple genera, species, subspecies and genetic sequences changing and evolving in association with people across multiple continents over thousands of years. Archaeobotanical evidence points to wheat being amongst the earliest domesticated plants in the world and the basis for one of agriculture's primary centres of development. In a sense, then, we have a classic example of the plant and the people domesticating each other, bringing one another into mutual relation.

Next, we consider the process of becoming *Australian* wheat. This is both a spatial process, involving global movements and transplantations, and a genetic one, as new Australian wheat varieties were bred in situ. All these processes were highly political; the lands that were to become the wheat belt had first to 'unbecome' something else, the Aboriginal grain belt. So, while this particular aspect of becoming provides important context for later chapters that focus on our Australian ethnography, it also stands as an example of the global politics of agricultural colonisation that occurred in many parts of the world.

Finally, we consider the way wheat became a *global commodity*, particularly in the period after the Second World War. Wheat has likely been a traded commodity almost as long as it has been a subsistence food, and, as the domestication story shows, it very quickly commenced its global travels. But the years after WWII saw an intensification of factors that have exacerbated and intensified these longstanding trends, with significant implications for food security and sustainability issues in the future.

These four long moments never definitively tell us what 'wheat' is, nor do they attempt to provide a story that is complete or whole. Rather they illustrate both the complexity and diversity of the wheat collective and how it comes into being in different places at different times. We also aim to illustrate that human-wheat relations are in an ongoing state of change as well as continuity; this becoming is simultaneously a spatial and temporal process.

Becoming Grass

Grasses constitute an ancient group of plants that originated in the Cretaceous and became widely distributed by the beginning of the Tertiary (Tzvelev 1989). Dating the chronology of evolution is more complicated for plants than animals due to a comparative lack of fossil evidence, and also because there are more complex histories of change and variation in plant DNA which challenge current dating methodologies (Kellogg 2001, Hedges 2002). Reliably dated fossil pollen grains of grasses are recorded from between 55 and 70 mya (Kellogg 2001). The capacities to tolerate drought and open environments are argued to be the most notable characteristics of grasses (Kellogg 2001), advantaging them over many non-flowering plants during the climatic aridification and cooling at the end of the Cretaceous (Tzvelev 1989). Kellogg (2001) argues that grasses had been a minor component of the world's vegetation communities for millions of years before they became ecologically dominant. Ecological expansion of the grasses took place sometime in the late Miocene (9–7 mya), dated using a range of techniques (Lunt et al. 2007). There is continuing debate about whether this expansion can be attributed to changing atmospheric concentrations of CO_2 or possible changes to seasonal precipitation during the period (Lunt et al. 2007).

Grasses, known botanically as Poaceae (alternatively Gramineae) and as monocots (having one cotyledon or seed leaf), make up the largest family of angiosperms (flowering plants). Some 900 genera and over 10,000 species are distributed across all continents and climatic zones of the earth (Tzvelev 1989). This is double the number of mammal species and equivalent to the number of all bird species on the earth (Kellogg 1998).

Research on faunal fossil assemblages and climatic modelling suggests that grasses played a critical role in the evolution of modern humans. A broad set of ideas, often referred to as 'the savanna hypothesis', suggests that hominid evolution and divergence from more tree-bound apes originated in the late Miocene as an outcome of broader climatic change. (In popular science this is often echoed in the idea that humans are genetically primed to view open grasslands as pleasing, for example Truett 2010.) Bobe and Behrensmeyer (2004) re-examined the savanna hypothesis, arguing that grasses might have been critical not necessarily because they stimulated hominid divergence, but because they provided habitat heterogeneity at a period of heightened climatic change. This stimulated high faunal (prey) turnover and increases in the abundance of common mammal species adapted to grasslands.

The methodological challenges of elucidating grassy evolution are complicated by the parallel scientific story of grass taxonomy, which began formally in the mid to late nineteenth century (Tzvelev 1989, Kellogg 1998). Like all systems of plant classification, grass systematics is generally arranged according to morphological characteristics of flower structure, with micromorphological features and genetic information having been introduced over time as the techniques were developed. The International Code of Botanical Nomenclature (ICBN) provides the guidelines and principles by which botanists classify and name plants; other organisms are classified according to their own codes (Singh 1999). The ICBN includes principles such as the naming of taxa, rules governing type methods, author citation and the publication and rejection of names. This governance of nomenclature has to deal with tensions arising from changing material practices, from field specimen identification to the use of genomic and molecular techniques that explore the full spectrum of phylogenetic diversity (Barkworth 2000). Wheat provides two important examples of the tensions between stable and unstable identities in this story; boundary making around the genera *Aegilops* and *Triticum*, and the conservation of species names for *Triticum aestivum*.

In the current system, the wheat genus *Triticum* belongs to the true grass subfamily (known as Pooideae, separated from the woody stemmed Bambusoideae cohort). Within Pooideae, Triticeae is the tribe to which both genera *Aegilops* and *Triticum* belong. This tribe encompasses a widely distributed group of tropical and extra-tropical mountain grasses, distinguished by their spike-like inflorescences[2] (Tzvelev 1989). Barkworth (2000) sketches the historical classification of Triticeae, showing how taxonomic treatments varied over time and contained major discrepancies between the classifications of *Aegilops* and *Triticum*. One example is the application of the hand lens by Russian botanists in the 1930s, which together with cytological and later genomic data, led to significant revision and expansion of this group of

2   An inflorescence is a cluster of flowers arranged on a branching stem.

grasses within the family. An organisation called the International Triticeae Mapping Initiative (www.wheat.pw.usda.gov/ITMI/) now exists to try and deal with the enormous complexity and challenges this grass tribe presents.

A further governing principle of the ICBN is the Principle of Priority, whereby if multiple correct names for the same species (known as synonyms) are legitimate according to the code, then the name with the earliest date of publication should take precedence. This principle has been applied to cases such as *Prunus dulcis* (almond) and *Malus pumila* (apple) without too much concern, but the discovery of multiple legitimate names for wheat caused great consternation and even protest amongst agricultural botanists and horticulturalists (Singh 1999). As Singh (1999) details, Linnaeus first published two species names, *Triticum aestivum* Linn., and *Triticum hybernum* Linn., in 1753. Following the principle, the first publication uniting these names as synonyms should take preference. According to Singh (1999), Hanelt and Schultze-Motel pointed out in 1983 that this had been done by Merat in 1821, when he selected *T. hybernum* L. (even though botanists and practitioners had been using *T. aestivum* for well over a century and a half by 1983). In response to this conundrum, taxonomists proposed a new rule – the Conservation of Names of Species – providing a critical limitation to the Principle of Priority. With this act, the 'number one economic species' (bread wheat), became *Triticum aestivum* Linn. at the International Botanical Congress in Berlin in 1987 (Singh 1999: 48). It is unlikely trivial that an exception had been made for wheat; wheat is not only taxonomically complex but also has more power to command human attention than other plants.

Becoming Domestic

In light of the critical debates around domestication that we outlined in Chapter 2, in what sense(s) do we understand wheat as becoming domestic? Morphological changes in the plants under human selection pressures captured archaeological attention, and for many years these modifications stood for domestication itself. Changes in morphology are important because they provide diagnostic markers of change and continuity, but the connections between the material change in the plant, and any associated social practice, need to be demonstrated rather than asserted. In the case of wheat practices such as cultivation and grinding, the evidence separates the practices in space and time; in the overview below we unpack the bundle of practices, following Denham (2007a, 2007b). Archaeologists increasingly recognise and discuss a broader understanding of domestication, encompassing broader changes in societies' relations with their environments. Attention to materiality and social practice reminds us also that, as with botanical taxonomy, our understanding of these issues is inseparable from the social scientific practice of archaeology. The history of wheat domestication continues to be refined over time with the development of modern genetics and DNA characterisation, and advances in archaeology and carbon dating techniques. Detailed accounts of wheat's domestication are provided in a number of texts (see Harlan 1981, Caligari and Brandham 1999, Zohary and Hopf 2000, Feldman 2001, Salamini et al. 2002). The scientific or botanical definition of domestication emphasises the physical (morphological) and genetic modifications between wild and domestic populations of plants made by humans selecting particular seeds from wild populations of plants and growing them over generations. These processes do not follow a simple trajectory from wild to domesticated as a lot of genetic diversity can be created through these evolving relationships, only some of which might become useful and selected for over time. In evolutionary terms humans exert reproductive control over a population by selecting traits, both consciously and unconsciously, in combination with random genetic events such as mutations or hybridisation events.

Human use of starchy grains has a much longer history with 'wild' than 'domestic' species. Recent evidence indicates that early *Homo sapiens* were processing sorghum grass seeds in Mozambique at least 105,000 years ago (Mercader 2009). Starch remains on grinding stones indicate that foragers along the present day shore of the Sea of Galilee were processing grass seeds, including barley and probably wheat, about 23,000 years ago (Piperno et al. 2004). This seed grinding was combined with the baking of dough at that time, at least 12,000 years before evidence of domestication.

The sites of wheat domestication are concentrated in what has become known as the Fertile Crescent, an arc extending from Greece and Turkey through to modern day Iraq and Iran. This region has many different habitats and plant communities, including a comparatively high diversity of wild annual grasses (Feldman 2001). Genetic evidence suggests that the initial introduction of a number of wild 'wheat' progenitors into human managed systems took place in a number of specific localities throughout the arc, the 'nuclear areas' (Zohary and Hopf 2000, Feldman 2001). Recent research narrows the core area to the upper reaches of the Tigris and Euphrates rivers, in what is today southeastern Turkey/northern Syria (Lev-Yadun et al. 2000).

An important distinction needs to be drawn between human alteration of the phenotype and the social practice of planting and harvesting seed for use (cultivation) (Salamini et al. 2002). Tanno and Willcox (2006: 1886) argue that 'wild cereals could have been cultivated for over one millennium before the emergence of domestic varieties'. Other archaeobotanical evidence from the Levantine Corridor clearly demonstrates that 'wild' wheats were harvested and subsequently cultivated by hunter-gatherers in the region between about 13,000 years and 7,500 BP, that is they were utilised extensively for quite a long period before any phenotypic or morphological changes identifiable as 'domestication' took place (Feldman 2001). Indeed wild wheat persists today as a weed in Turkey.

The classic morphological marker of wheat domestication is indehiscence (Figure 3.1). As Tanno and Willcox (2006: 1886) explain,

> Wild cereals with dehiscent ears shatter at maturity into dispersal units called spikelets, identifiable by their smooth abscission scars ... The first domestic

cereals arose from mutants, which have indehiscent ears with spikelets that do not shatter but separate when threshed, identifiable by jagged scars.

This indehiscence facilitated harvesting and threshing of the wheat in a contained way, constraining the spikelets against their evolutionary duty to disperse and grow elsewhere. Imagining the everyday practice of threshing leads to the inference that 'the non-brittle types were probably selected by women, who were usually in charge of threshing' (Feldman 2001: 49), both the wheat and the women becoming less mobile over time. On these criteria the earliest indehiscent domestic wheat is dated to about 9,250 BP in the region of southeastern Turkey and northern Syria (Tanno and Willcox 2006). Other diagnostic characteristics include erect types, synchronous tillering and uniform ripening; increased seed production; reduction in awns, and glume thickness (Zohary and Hopf 2000) (Figure 3.2). In the absence of human intervention, evolutionary processes would select for very different traits such as non-synchronous tillering, which would increase the success of progeny under non-cultivated environments.

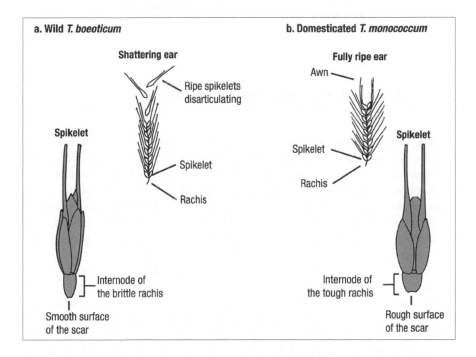

Figure 3.1 Morphological differences between wild and domesticated wheat with relevant terminology

Source: Adapted from Salamini et al. 2002: 431.

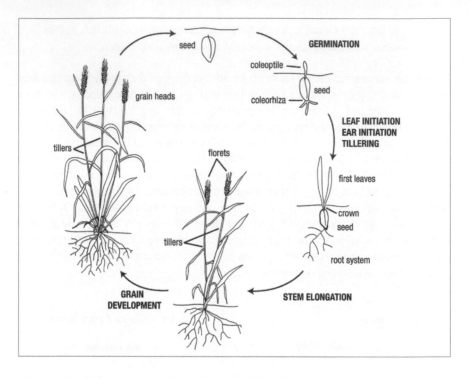

Figure 3.2 Life cycle of a domesticated wheat plant
Source: Adapted from Perkins 1997: 21, original by Kirby and Appleyard 1987: 4.

Amongst the great variety of domesticated wheats which emerged from the near east arc, the appearance of hexaploid wheat, *Triticum aestivum*, a species for which there are no wild progenitors, is an especially significant moment in this story. This species, which now represents about 90 per cent of the contemporary global crop, is thought to have originated in a hybridisation event between a tetraploid turgidum wheat (an earlier domesticated wheat) and the wild grass species *Aegilops squarrosa*, a weed and coloniser of opened cultivated spaces (Zohary and Hopf 2000), when tetraploid wheats were transmitted into the geographic range of *A. squarrosa*. While the event most likely took place somewhere in the south-west corner of the south-western Caspian Sea belt (Zohary and Hopf 2000), there is some disagreement about exactly when hybridisation took place. The archaeobotanical evidence predates and contradicts (by about 1000 years) other archaeological and biological evidence that it could not have taken place until after domesticated tetraploid wheats were first cultivated in the Caspian Sea after 6,000 uncal. BCE (Nesbitt 2001). Whatever the exact timing, hybridisation both significantly extended the potential environmental and geographic range of the turgidum wheats and concurrently resulted in a cereal crop with enormous genomic variety; both factors underpin its contemporary success.

By Feldman's (2001) account, most of the major evolutionary steps toward the full contemporary suite of domesticated wheats known today had been completed by about 7500 BP. Dating of remains suggests that the domestication of wheats such as einkorn and emmer took place approximately 7500 years ago (Nesbitt 2001). These domesticated grains were quite visually distinct morphologically as non-brittle and 'naked' forms. This relatively rapid burst of diversity is argued by Feldman (2001) to be the result of a range of new environmental conditions and selection pressures provided by cultivation which were not present in wild conditions. However, domestication does not 'stop' at this point. A series of ongoing mutations are evident, including the formation of free-threshing bread wheat, the form with which we are familiar today (Table 3.1).

Table 3.1 Morphological modifications to wheat during phases of domestication

I During the transition from the wild habitat to the cultivated field. (* = main traits that conferred adaptability to agricultural conditions)	1* Non-brittle spikes 2* Free-threshing (naked grains) 3* Non-dormant seeds 4* Uniform and rapid germination 5 Erect plants 6 Increased grain size 7 Increased spikelet number per spike
II During 10,000 years of cultivation in polymorphic fields. (* = main traits that increased competitiveness in poly-genotypic stands)	1* Adaptation to new, sometimes extreme, regional environments 2* Increased tillering 3* Increased plant height 4* Development of canopy with wide horizontal leaves 5* Increased competitiveness with other wheat genotypes and weeds 6 Modifications in processes that control the timing of various growth stages 7 Increased grain number per spikelet 8 Improved seed retention (non-shattering) 9 Improved technological properties of grains
III During cultivation in monomorphic fields due to modern breeding procedures in the last century. (* = main traits that reduced competitiveness in dense mono-genotypic stands)	1* Increased yield in densely planted fields; reduced intragenotypic competition 2* Canopy with erect leaves 3 Reduced height 4* Enhanced response to fertilisers and chemicals 5 Increased resistance to grain shattering 6 Increased resistance to diseases and pests 7 Lodging resistance 8 Improved harvest index 9 Improved baking and bread-making quality

Source: Adapted from Feldman 2001: 28.

The early domesticated wheats were transmitted extensively along trade routes throughout Africa, Europe and Asia (Figure 3.3). There is some debate about the first wheat to have arrived in Europe. Feldman (2001) argues that wheat arrived in two main waves, the first at about the same time as the first traces of agricultural activity. Salamini et al. (2002) argue that a diploid variety known as Einkorn (*Triticum monococcum*) arrived in central Europe as early as 7000 cal. BP, although Feldman (2001) suggests forms of hexaploid wheat came in as an 'admixture' with both Einkorn and Emmer. In this history it is important to remember that wheat has many uses other than human food. Egyptian use of wheat starch to help produce a smooth surface for papyrus documents was described by Pliny the Elder in the first few decades of the Christian Era (CE). This has recently been confirmed archaeologically with the dating of papyrus, bonded with starchy adhesive, to 3500–4000 BCE (Whistler and Daniel 2000).

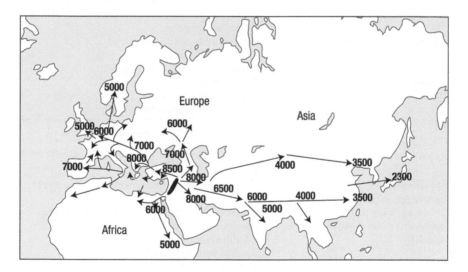

Figure 3.3 Pathways of wheat expansion into Europe, Asia and Africa
Source: Feldman 2001: 28.

This overview of domestication gives little sense of the human dimensions; the untold hours of careful observation of when, where and how plants grow; the labour of harvesting, threshing, winnowing, grinding and cooking many different grains; the deliberate and accidental experiments at all stages of the life cycle. Storable seasonal surpluses offered new possibilities and choices in the scheduling, location and concentration of social life. They also required attention if they were to be protected from the weather, pests and other human demands. Investment of time and effort in the high returns promised by cereals entailed risks if the crop failed, and the evidence suggests that in most early agricultural contexts, hunted or gathered food continued to be important. Nor should we forget that for most of its human history, wheat has had a variable history as good food. Knobloch (1996: 52) for example quotes Braudel and others about wheat's chronically low yield as food, and the necessity of women gardening other vegetables in order to avoid famine in Europe as late as 1400–1800 CE. There are several reasons why it is difficult to enrich the story with a sense of everyday life and labour; the most obvious is that prehistory is lost to us except insofar as it can be approached via archaeological research. Archaeologically, wheat has been part of the morphocentric cereal domestication stories subject to critique from other parts of the world where genetic modification was less relevant to the appearance of agriculture, for example Papua New Guinea, as discussed in Chapter 2.

We can achieve a more social understanding in two particular ways; by ethnographic analogy and through a closer examination of the material culture associated with wheat production. Ethnographic analogy is itself fraught with unjustified assumptions if hunter-gatherer use of grains is interpreted as being 'on the way' to agriculture, but it remains an important window onto practices that are themselves variable in space and time. The description of Australian Aboriginal grain use in the following section provides one such example. The material culture of grindstones, knives, ploughs, hoes, scythes and storage vessels reminds us that the relationship is never one between just the people and the plant, but that there are many other participants in the network. Today archaeologists use microscopic starch residues, squelched into tiny cracks in the grindstone, to help elucidate relationships between people and plants from thousands of years ago (Fullagar et al. 2008). Important actors in the Australian part of the story include desert sandstones from which grindstones were quarried, a quarryman who marked his slabs with the outline of a little quail, knowledgeable old women, and pools of available water. Northern hemisphere regions have their own variations.

Becoming Australian

Skipping a few thousand years forward in time, we can consider the further globalisation of wheat under the process of European colonisation. This was not only a shift of locations, but also a shift from the wet and mild conditions of Europe to hotter, colder and drier centres of production in the Americas, Australasia and South Africa (Olmstead and Rhode 2007). Wheat was famously on the cargo list of Australia's first fleet, which brought convicts to found Sydney in 1788. The England to Sydney leg was just another link in the chain of wheat transplantation that had been occurring for thousands of years. The process of becoming Australian wheat was not only a spatial one, important though that was. There are many practices through which Australian farmers and plant breeders have been in a continuous process of adapting wheat to the particular circumstances in which they find themselves. These should be understood as part of a broader evolutionary history in which human involvement with, and movement of, plants is an integral part, not an aberration.

Unbecoming the Aboriginal 'grain belt'

Before what we now call the Australian 'wheat belt' came into being it had to unbecome something else. Pastoral and agricultural expansion in New South Wales
(NSW) was a process of dispossession of the Aboriginal owners. Many aspects of
that dispossession have been commented on in more detail by others. Here we simply
note that part of what was lost was a bundle of practices referred to by anthropologist
Norman Tindale (1974) as the Aboriginal 'grain belt'. The boundaries on Tindale's
map (Figure 3.4) should not be taken too literally – they show the imprint of where
the 1970s Letraset was cut – but there is clear overlap between the southeastern end
of his grain belt and the current NSW wheat belt, 'in the flatlands west of the Great
Dividing Range' (Tindale 1977: 345). (Nor should the concept of a grain belt imply
a social unity, beyond the coincidence of practices since, as Tindale noted, these
widely scattered peoples differed widely in the details of their social organisations
and ceremonial practices.) Nineteen seventies conventional wisdom held that
Australian Aboriginal people, like other hunter-gatherers, existed in harmony or
balance with nature and did not intervene in its processes. Ethnographic evidence
of Aboriginal practices around seed grinding were sufficient however to provoke
comment, and Tindale was tempted to speculate that this could be evidence of the
'Pre-Dawn of Agriculture in Australia':

> There is evidence for the incipient development of grain storage, and in two
> situations a suggestion that the effectiveness of seasonal flooding of grassed
> plains had been modified by damming stream beds prior to the advent of summer
> rains, implying incipient interest in irrigation. (Tindale 1977: 345)

Use of many different seeds has been recorded. Mulvaney and Kamminga (1999:
85) note the importance of wild millet (*Panicum decompositum*) and the sedge
Fimbristylis oxystachya along the Darling River. The journals of early European
explorers in this region were also important sources of information (Mitchell
1848, Sturt 1849). For example:

> In the neighbourhood of our camp the grass had been pulled, to a very great
> extent, and piled in hay-ricks, so that the aspect of the desert was softened into
> the agreeable semblance of a hay-field. ... when we found the ricks, or hay-
> cocks, extending for miles, we were quite at a loss to understand why they had
> been made. All the grass was of one kind a new species of *Panicum* ... and not a
> spike of it was left in the soil. (Mitchell 1848: 237–8)

Tindale summarised the 1905 writings of Mrs Langloh Parker regarding
agricultural practices among the Ualarai of the Walgett area on the Upper Darling:

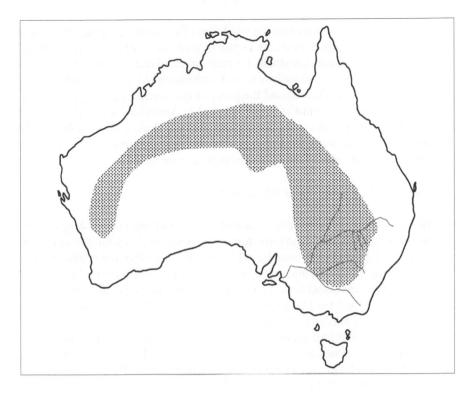

Figure 3.4 The Aboriginal grain belt
Source: Mulvaney and Kamminga 1999: 86, after Tindale 1974.

> In that matrilineally oriented society men played an important part in the
> harvesting of the grass seed and ... the product, which [Mrs Parker] called
> *yammara*, was stored against future use. Women collected the ripening ears of
> grass while still green and they were piled into a brush enclosure and the whole
> set on fire. Women turned the pile to shake out the parched seeds, using long
> sticks. The seeds were then piled on possum skin rugs. Men took the part of
> removing the husks by treading them in a square hole in the ground. Other men
> worked a stick around in a circular hole filled with the trampled grain, causing
> the husks to work their way to the top. Further winnowing and the use of bark
> dishes, known as *wiri* and an especially large canoe-shaped bark vessel known as
> a *jubbil* completed the cleaning of the grain. Since there was a harvesting season
> the grain was stored in skin bags until required. Then the grain was prepared for
> eating by wet-grinding on millstones called dajurl and made into flat cakes to be
> cooked in the ashes of a fire. (Tindale 1977: 346)

In the diverse examples referred to by Tindale and other writers, both men and
women are involved in different parts of the process, but the grinding of seeds

seems to have always been women's work. Indeed Tindale quotes an (unreferenced) 'women's lament which protests the inertia of men who will not make the travel effort necessary to obtain new stones for their wives' (Tindale 1977: 347).

The combination of ethnohistoric and archaeological evidence attests to a long continuity of practices around the processing of seeds and other plant parts. The tenure of the seed grinders is still marked in Australia's arid and semi-arid landscapes by the ubiquitous presence of grindstones. These could weigh up to 20 kg and were often cached under trees or in rockshelters for return use (Fullagar et al. 2008), a use cut short by the invasive landscapes of wheat.

Becoming the New South Wales Wheat Belt

By the time of Sturt's and Mitchell's journeys, early farmers in the colony of New South Wales were attempting to grow foods they were familiar with (the edict from London was the penal settlement be self-sufficient), but from the beginning they were having to do so in quite unfamiliar circumstances. Before the 1870s NSW wheat was grown as a subsistence crop in coastal regions such as the Hawkesbury-Nepean lowlands and the Hunter Valley, on very small 'blocks' of not more than about 50 acres (Robinson 1976: 162). There was also isolated production on the Goulburn Plains, Bathurst plains, and in the Illawarra/Shoalhaven, near Nowra. Few people were trained or had much experience in agriculture, and there were few stock available for agricultural labour, which was initially performed by convicts (Dunsdorfs 1956).

During this 'foundation' period there was no economic market for growers to sell into. Robinson (1976: 20) detailed correspondence between Governor Hunter and the Colonial Office in London about why the absence of a market for grain was problematic. London's view was that as the 'purpose' of NSW was a penal colony, it was not part of the government's role to support private farming. Wheat was thus part of initial attempts to frame an Australian identity distinct and separate from England; it was 'of great importance in the conversion from a penal settlement into a colony' (Dunsdorfs 1956: preface). Two thirds of farmers in this period were ex-convicts, as this was the only occupation they could readily take up without capital.

Important land tenure changes occurred between the 1850s and 1870s. Following a Legislative Council enquiry (1855) into the lack of agricultural (specifically wheat growing) development, tenure reforms were created to deal with uncontrolled pastoral expansion and squatting. The original Order-in-Council land subdivision of 1847 had given squatters the right to lease land in intermediate and settled districts, but prohibited the use of leasehold land for commercial agriculture, placing pastoral interests at an obvious advantage in inland districts. Robinson argued that 'the rising tide of liberalism during the 1850s demanded reform of the whole structure of land settlement regulations, and the outcome of these demands played a significant role in developing the wheat industry' (1976: 51). Reform came in the form of the Robertson Land Acts of 1861, widely criticised and debated in the historical literature as a large amount of public pastoral land passed into freehold tenure. However valid these criticisms, Robinson argued that the legislation resulted in fundamental changes in the distribution of settlement to the inland. There were increases in the land area put to agricultural production, and pastoralists were able to diversify their enterprises, to both run sheep and grow wheat. There were differences between the colonies in their wheat farming from the start. For example, South Australia produced more surplus for export than New South Wales and, with its more Mediterranean climate, was less susceptible to rust and other humidity diseases. Indeed Sydney and surrounds initially were regularly dependent on 'imported wheat' from South Australia, Tasmania and from other countries such as India.

Robinson (1976) argued that there were three main factors in the shift of NSW wheat from coast to inland. Or rather, the wheat did not shift but the inland became established before the coast crashed as a centre of production. Disparate inland centres filled in and subsequently expanded, rather than there being a westward expansion of a farming frontier. First, drought in the 1850s contributed to partial and total harvest failure across the coastal districts, and there were widespread shortages in grain and flour. South Australia's wheat crop had been sold to South Africa, and New Zealand did not have any wheat to export. This situation led to a dramatic increase in the price of wheat. Paradoxically, from the earliest years, this interplay between price, scarcity and demand made wheat an attractive proposition in ongoing cycles of drought. Second, the 1850s gold rush contributed to the development of wheat inland, both to feed the miners, and also by providing a labour force as miners left mines to find more permanent employment. With farmers able to get good prices, acreages grew and regular surpluses became available, although the inland and coastal markets essentially operated separately. Third, stem rust (*Puccinia graminis* f.sp. *tritici*) contributed to the collapse of the more humid coastal wheat districts. Rust was familiar to Australian farmers very early on, but it was not endemic before the 1860s. In Australia, unlike Europe, the fungus does not pass through a vulnerable stage during winter, thus the spores survive in wheat stubble and in 'volunteers' (plants which sprout in a field from unharvested or spilt grain).

The first half of the 1860s saw a 60 per cent reduction in coastal wheat acreages in less than five years in. By the end of the 1860s the Orange, Bathurst and Yass districts were the only districts where a significant amount of wheat was grown, and in the 1870s these were overtaken by places further west. These major geographic changes barely register on the historical wheat inventory, because the inland districts so rapidly expanded and 'took over' the production reins. Subsequent expansion inland was still not possible until the full development of the rail network. Before the rail,

> production outstripped the resources of local carriers … the scarcity of team haulage meant that … costs ate deeply into profit margins of wheat farmers …

slowness and uncertainty delayed the response of market fluctuations causing
serious losses when markets were high. (Robinson 1976: 209)

The initial rail trunk development phase, finished in the early 1880s, developed
around providing supplies to the inland and the transport of wool back to the coast
(Figure 3.5). A significant expansion phase followed after 1888, the major objective
being to secure the NSW wool trade so farmers did not 'export' to Melbourne or
Adelaide. After initial development, branch expansions were selected on basis of
regional profitability, particularly the provision of feeder lines in southern and
central regions. After the 1890s the rail expansion policy was developed for wheat.

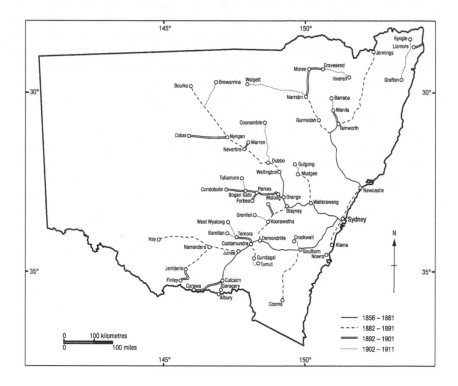

**Figure 3.5 Major phases in the expansion of the NSW regional rail network
1865–1911**
Source: Robinson 1976: 194.

Two other innovations in the late nineteenth century were considered by Robinson
to mark a final break with British agriculture. Technological innovations in
Australia included new plough technologies such as the stump jump plough, which

could cope with tree roots in paddocks under cultivation, and new harvesters and threshers better suited for growing wheat in dry environments (Dunsdorfs 1956: 149). These 'home grown' inventions from South Australia (Callaghan and Millington 1956) were not immediately taken up in NSW. Second, Australian farmers were in conversation with other dryland farmers in colonial and former colonial countries. For example the system known as dry-farming (or the Campbell system), whereby the field is ploughed after harvest to provide a 'dust-mulch', was brought in from the US. New ploughs such as the disc plough were developed to break through hard arid soil crusts. Through the 1890s and 1900s the new machinery and technology of dry farming, which enabled the expansion of cropping areas into drier more arid environments, 'focused attention on the problems of tillage and moisture conservation' (Robinson 1976: 176).

So in this example, two things which are often depicted as the bane of the Australian wheat industry – drought and rust – had considerable agency in its survival and expansion. We are very used to thinking of drought as bad for wheat, but that is only partly true. Drought continues to be significant in maintaining demand for wheat, and it always guarantees good prices. We also need to think of the different scales across which drought pulses. In the season of writing (2010/11), Australian wheat farmers, expecting a bumper drought-breaking harvest themselves, have been quietly pleased about the drought in Russia, as it has driven up demand for their own wheat.

Against the Grain: Aboriginal Connections in the Wheatlands

Knobloch (1996) depicted the American wheatlands as an invading state. This sort of totalising vision of the colonial enterprise has required revision in the last couple of decades, giving way to a more nuanced view of colonialism in which indigenous people have some agency. In the Australian context this is certainly now the understanding of the pastoral frontier. A large literature attests not only to the importance of Aboriginal labour as part of that industry, but also the compatibility between the seasonal nomadism of both arid zone pastoralism and traditional Aboriginal lifeways (McGrath 1987, Goodall 1996, Gill and Paterson 2007). Goodall (1996) called this 'dual occupation', and Hannah (2002: 17) has argued 'that there was an accord between work rhythms in subsistence economies and the attributes required of pastoral workers in the early colonial period.' Perhaps because grain cropping and its fences blanket and mark the landscape in more static ways than cattle and sheep pastoralism, we have presumed that agriculture is a stronger excluder, as articulated by Knobloch: 'The western bonanza farm is an example not of colonization gone wrong but of colonization in its most literal manifestation: enclosing property for the cultivation of commodities' (1996: 57).

Recent empirical work that gives voice to Aboriginal experience is disrupting this view. Bennett (2005) draws on the detailed agricultural records of Alexander Berry's Coolangatta Estate, at the mouth of the Shoalhaven River near present-day Nowra (Figure 3.5). Dating to the 1820s, this 10,000 acre land grant was part of the 'foundation' period of wheat growing in Australia. Bennett argues for a different sort of dual occupation than that used by Goodall for pastoral properties. In the early years most Aboriginal people maintained traditional hunting and gathering subsistence patterns, occasionally assisting on the estate, particularly when there were problems with convict labour. Although initial assessments of Aboriginal harvesting techniques were less than favourable, Bennett reports that 'their skills improved and Aboriginal people continued to assist with corn and wheat harvesting over many decades' (2005: 4). During the 1830s the tasks Aboriginal people undertook included a combination of traditional bush and new agricultural skills, the latter becoming more important over the following decades: 'collecting bark, fishing, boating, tracking horses, capturing convicts, delivering messages, sewing and reaping crops, threshing seed, washing sheep, making yeast, washing bags and cleaning the storehouse' (2005: 4). Bennett reports that

> by the 1880s, Aboriginal people of the estate did not necessarily have to work
> or hunt and gather in order to obtain subsistence. Following the creation of the
> Aborigines Protection Board in 1883, rations were available to children and the
> elderly, reducing the pressure on younger adults to provide for family members
> incapable of supporting themselves. (2005: 9)

Like Aboriginal people across Australia, people at Coolangatta would in fact have been less able to hunt and gather, as habitats for animal and plants were drastically changed, although some traditional modes of subsistence such as fishing persist today. Late nineteenth century rations of course included flour, sometimes of dubious quality. Today the Coolangatta area and the banks of the Shoalhaven are a wheat place of a different type, being the location of the starch plant discussed in Chapter 8.

To come closer to the present, McCann (2005) recorded stories by Wiradjuri people in the Lachlan Valley who worked on local farms, and by white farmers who grew up with the children of those workers. Barbara Allen, who grew up at the Condobolin Mission, and later did a full range of agricultural labouring, said of the farmers, 'because they love the land and [my] people love the land, you feel so close to them' (McCann 2005: 10). The material remains of previous Aboriginal occupations also provide resistance against the plough; 'One farmer recalls the thrill of excitement as a child, always digging up axe heads and polished stones on his family's farm, and taking them to show his father' (McCann 2005: 11). In a context of agricultural decline since the post WWII heyday, it is paradoxically the white farmers who are grieving for the loss of community and attachment to place:

> Ironically, whilst the settler stories of old places are infused with nostalgia and
> regret for eroded values and lost community, Wiradjuri stories convey a sense
> of vitality and immediacy in the present. The Wiradjuri experience suggests
> that stories of connection still move lightly amongst the orderly grid of fenced
> paddocks. In the central Lachlan, Wiradjuri history is literally written into the

past. But, all the time, the act of remembering enables Wiradjuri descendants to work against the grain of dominant settler narratives. (McCann 2005: 12)

Breeding 'Australian' Wheat

The lineage of early wheat varieties brought to Australia is hard to trace because little effort was made to describe lines or keep them true. Early varieties are now understood to be derived from collections made en route from England in Rio de Janeiro, through which diverse lines were unwittingly introduced to Australia (Aitken 1996 in Olmstead and Rhode 2007: 126). After the 1850s White Essex introduced from Britain was successful and commonly grown (Wrigley and Rathjen 1981: 98), but the process of wheat breeding started early. Analysts have tended to focus on the discontinuities in this story; the foreignness of wheat and agriculture to Australian soils and climatic conditions, focusing on the differences between England and Australia.

> ... more than in the countries of the New World, the wheat breeder has had a unique part to play [in Australia]. For here he had the task of introducing and adapting a completely foreign plant to a continent with no known tradition of agriculture as well as having been essentially isolated from other land masses since well before the domestication of wheat. (Wrigley and Rathjen 1981: 96)

In this section we also want to highlight some continuities. Much of wheat's evolutionary story, including its grassiness, has occurred in climatic conditions not dissimilar to the parts of Australia where it is now dominant. And, although the timescales are different, the vernacular experiments of getting wheat to 'belong' in Europe from its semi-arid middle eastern origins would have been just as complicated and fraught with failure as those involved in making wheat Australian. The England-Australia link is just one part of a breeding story that was always more international than usually narrated, with early connections between Australia and India, Mexico and Argentina among others.

The selfing characteristic that was such an important feature of wheat's early domestication can become a problem under cultivation if a population is isolated or unable to outcross with wild populations. It takes a very long time for natural hybridisation to produce variety. The breakthrough of artificial pollination meant that people were able to cross-pollinate plants directly, speeding up the process. The first Australian experiments on wheat were undertaken at Roseworthy, South Australia in 1882, around the same time as experiments in the US and Europe (Wrigley and Rathjen 1981: 104). The effort was spurred on by calamitous crop losses due to a rust outbreak in 1889, when over £2.5 million was lost (Wrigley and Rathjen 1981: 106, compiled from Campbell 1911 and Waterhouse 1936). The last decade of the nineteenth century was a significant period for the introduction of new wheat lines into Australia. Both imports and breeding efforts focused on quality and the plant's ability to grow in dry and semi-arid environments, away

from the humid rust-prone coastal areas. Pedigrees summarising the lineages for
well-known varieties such as Federation illustrate this multicultural heritage, with
genetic input from Italy, Scotland, South Africa, Canada, Hungary, India and US
(O'Brien et al 2001: 618–20) (Figure 3.6).

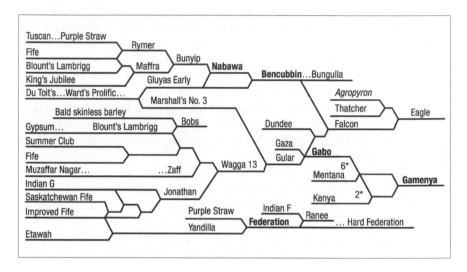

**Figure 3.6 Illustration of an Australian wheat pedigree of the varieties
Federation, Nabawa, Bencubbin, Gabo and Gamenya, including
introduced source material from India, North America and Europe**
Source: O'Brien et al. 2001: 620.

Federation is undoubtedly the most famous of the new 'Australian' wheats, and
the breeding work by William Farrer is one of the best known wheat stories
(Robinson 1976, Wrigley and Rathjen 1981, Lupton 1987, Hare 2001). This
'Australian' story is already an international story, since imported wheat lines and
later germplasm were critical, and a suite of imported genetic material was finetuned to local environmental conditions. Farrer was excited by the potential of the
breeding process – in the early years of his breeding experiments he was publicly
criticised by the sceptical media. With very little formal training, he saw immense
potential in creating – 'making' (Robinson 1976: 187) – different kinds of wheats
for different purposes. As well as breeding for environmental conditions, Farrer
and other breeders of the time developed and finetuned the breeding process to
accommodate a range of different end uses, such as softer wheats for biscuit
making, and harder higher protein wheats for bread making. By 1906 Federation
had become well established (Robinson 1976: 191), and by 1911, NSW was the
most significant wheat producer in the colonies (which had by then become states).

Since the 1960s there has been a significant increase in the number of varieties potentially available to farmers. The new varieties from this period, considered to be the most important phase of germplasm introduction since the 1890s, out-yielded all previous Australian varieties (Wrigley and Rathjen 1981: 128). Alien genes (not from *T. aestivum*) continue to be significant in modern wheat breeding. Sears lists some of the wild relatives of wheat which 'are accessible sources of genes for use in wheat improvement' (1981: 75). 'Alien' germplasm was significant, for example during the green revolution when European germplasm, derived from *Aegilops ventricosa* and *Triticum persicum*, was brought into Australia in the 1960s for disease resistance. *Agropyron* has also been used for disease resistance, as has rye. In the late 1980s germplasm from *Aegilops tauschii* (previously known as *A. squarrosa*, the species thought to be the original hybridising cross) was brought in again for disease resistance and continues to be significant as 'a direct and fast mechanism for introducing new variation' (O'Brien et al. 2001: 619). Wheat breeding programs also produce what are known as 'synthetic' lines, for example crosses between *Durum* and *Aegilops*, as sources of new genetic variation.

What do Wheat Breeders Do?

The complexity of the breeding task is illustrated by wheat scientists who continue that tradition. We interviewed two research scientists working as 'wheat breeders', in December 2006. Although they have a set of generic plant skills and training in pathology, molecular biology and agronomy, the genetics of wheat are complex enough, and the process of breeding new varieties long enough, that scientists usually stick with this particular crop throughout their careers. From beginning, through selection, testing and to the final release, the production of a new variety can take 10 to 12 years.

In very simplified terms, wheat breeding is a stepwise process made somewhat easier by the fact that wheat is self-pollinating. The first stage involves identifying shortcomings in existing or current varieties, for example in crop yield or disease resistance. Second, parents with desirable traits, or possessing genes which when crossed together will produce offspring with desired traits, must be identified. Once those parents are identified they are grown in either glass houses or in plots outside and then crossed to form what is known as a segregated generation. That segregated generation is grown in a bird proof enclosure and tested for the particular trait or group of traits, and selections are made from the segregated population that are genetically stable. At this stage, the progeny of the genetically stable segregated generation, known as F2, are again tested against a number of other evaluation criteria.

Later F3 generations are sown in small trial plots, normally about one and a half hectares, on a number of field stations throughout NSW. At the F4 generation, the seed is harvested from disease resistant plants and then sown into yield trials at separate locations. Some disease susceptibility is tested in disease nurseries in regional locations, but susceptibility to aggressive or exotic rusts and other pathogens is tested at the University of Sydney's Cobbity research station, outside the NSW wheat belt, in an attempt to quarantine the commercial growing areas and maintain biosecurity.

By the sixth generation a test variety has been sown and tested in multiple locations for two years and at this point true breeding families can be selected. If successful these are then sent for further yield trials, and after this, varieties enter what is known as the National Variety Trial, the next series of larger scale multiple location trails confirming disease resistance, acid or saline soil performance, baking performance criteria and so on. Like normal wheat growing, breeding and variety trials are subject to varying environmental conditions including drought. Recently a particularly virulent Western Australian pathotype of stripe rust impacted not only commercial production, requiring additional crop spraying of fungicides, but also breeding programs across Australia, with some places losing up to 70 per cent of breeding stock.

Terms like 'drought' or 'disease' resistance are sometimes used to generalise about what is being selected for in the process of wheat breeding, but these terms belie the extraordinary complexity of selecting for genes across a hexaploid complex originally derived from at least three separate plants, in one of the most genetically complex organisms on earth. As the two wheat breeders explained to us, the complexity of drought resistance, for example, is mind-boggling.

> There have been examples of people who've said well great Eureka, we've solved the problem of drought resistance, and in all cases they've been nice pieces of science which have actually demonstrated that in some, an individual gene contributed to drought tolerance … the reality is that high levels of drought tolerance are not just those two things alone, it's those two things plus a whole range of other things added in.

> In some environments drought resistance, or drought tolerance is no use unless you have acid soils tolerance combined with it, because the roots just can't grow in the acid soils. In other environments it's not good unless you've got tolerance to high levels of boron because in alkaline soils, some alkaline soils, boron levels are toxic.

> So drought tolerance as such is an environmentally determined trait in a lot of cases, so you can either have pure tolerance to low water or soil water … But in a lot of cases drought tolerance has improved by just ignoring the drought tolerance completely and just selecting for tolerance to soil problems like acidity or boron. (Wheat breeder, NSW)

In reality drought tolerance is often achieved by avoiding the period when moisture stress on the plant is greatest. There are a number of notable varieties that give the farmer the option of doing that. For example, both H45 and Ventura are early maturing varieties that finish their developmental cycle earlier, avoiding the heavy water requirements and high moisture stress commonly experienced during the spring. However the suitability of these varieties is regionally specific; early maturing varieties are not suitable in environments which experience early frosts.

Once a variety is released, a new process of vernacular testing outside the laboratory commences, usually driven by the more innovative farmers in a district. Trial co-operators are key agents in the rapid dissemination of variety information. Quick uptake of new varieties can have advantages, but the more conservative approach can also pay off. As Fred, a government cereals extension expert, described to us, the recent introduction of an early sowing variety, a potentially really important advance, in fact proved disastrous when unexpectedly severe frosts arrived.

> I think 1998 was probably a good example there, that even though we had a spread in our sowing time we had early sown varieties through to quick sowing varieties. The growers that really got hurt by that were the ones that had the huge machinery and sowed their crop in a quick period of time and then we had that really late frost that we've never seen before and it basically took out our whole production. Whereas the growers that were probably more conservative, the older machinery and couldn't get their crop in on time and had a much wider spread in their sowing were probably less affected. They lost some of their crop but not total crop. (Fred)

Australia is currently in the midst of a shift from public to private breeding programs. This change has significant ramifications, among them the question of plant variety rights. At the time of our interviews with wheat breeders and government agronomists there was considerable uncertainty regarding funding and coordination of the new market based system, and also the impact the commercialisation process would have on the National Variety Evaluation Project. This project evaluates wheat lines (and other cereals and pulses) in trials from all public and private breeding programs across Australia (O'Brien et al. 2001). The National Evaluation Trials are funded by the Grains Research and Development Corporation through grower levies. Growers are also required to pay end point royalties – a fee to breeding programs that have developed a successful wheat variety upon delivery of a crop of that variety into the market. Currently there is no quick method for evaluating or identifying wheat varieties at the point of delivery, and some growers, disgruntled about paying both levies and royalties, have been known to deliberately misname varieties so that they do not have to pay end point royalties.

The process of wheat breeding and developing new varieties involves continuously adapting to changing environmental, disease and even market conditions. Contemporary processes share both continuities and differences with the earliest domestication and global movements of wheat. Pervading all of these is the mobile identity of wheat.

Wheat Becomes a Global Commodity: Food, Feed and Aid

The years following World War II saw transformative changes globally, some of them fuelled by industrial changes from war research. These included the massive increases in yield and productivity of cereals during the so-called Green Revolution between approximately 1945 and 1970, stimulated by breeding improvements and 'artificial' fertilisers (Lupton 1987, Reynolds and Borlaug 2006). There was also a major geographical reconfiguration of cereal, particularly wheat, movement across the globe. Before the Second World War grain historically 'flowed' from the global south to the global north. Since the Green Revolution the majority of that flow has been reversed, with developing countries importing grain from developed countries. (These trends are discussed further in Chapter 4.) Grain provided cash flow for the world economy after WWII. Recovering from the war, Europe could not feed itself, and looked to America to provide the wheat. The US supplied half the world wheat trade from 1945–49 (Morgan 1980: 136), having overtaken Australia as the most significant exporter by volume around 1930 (Broehl 1992: 573). This is not only a spatial becoming, but it is also a significant geopolitical one, a key component of which is wheat's shift from human food to animal feed.

The term Green Revolution (GR) is used generally in reference to significant yield improvements in a number of crops including maize and rice. However, Perkins (1997) argues that the first breakthroughs in breeding, and eventually the largest changes in yield, were seen in wheat. Wheat was also a major crop in the four countries where significant plant breeding research first took place; UK, US, Mexico and India (Perkins 1997). Newly developed scientific relationships between plant breeding programs saw breeding materials – germplasm – relocated and moved amongst a combination of new countries.

The success of the new varieties was underpinned by massive inputs of synthetic fertilisers and pesticides, together with new irrigation and dam infrastructure. Perkins (1997) summarises the GR as a simple equation of seeds plus nitrogen plus water (see also Leigh 2004). The conditions enabling the GR were also the basis of its environmental critique as an unsustainable type of agricultural production operating outside natural limits (Shiva 1993).

Semi-dwarfing varieties of wheat had been in use in Europe for some time, and were known to be especially useful where artificial fertiliser was applied. Longer strawed varieties of wheat had the propensity to fall over at harvest time when weighed down with grain, making mechanical harvest difficult. Shorter strawed varieties more efficiently made nitrogen into grain rather than into long vegetative stalks, and also held the wheat heads up through to harvest. The most successful breeding of short varieties occurred from semi-dwarfing wheats first identified in post war occupied Japan (Lupton 1987). Plant material was transferred back to the United States, further investigated by the International Centre for Wheat and Maize Improvement (CIMMYT) in Mexico and subsequently transferred, tested and bred in India and elsewhere. The transfer of these new wheat varieties into lower latitude (and also less developed) countries was also facilitated by a second major change in plant architecture, the elimination of the photoperiod response which, in more tropical environments with longer nights, results in lower yield (Trethowan et al. 2007). The origin of the germplasm used to breed these changes is unknown but possibly derived from Australian material (Trethowan et al. 2007). The new semi-dwarfing varieties grown under the 'correct' environmental conditions transformed wheat yields – in the space of two to three decades crop yields in developed countries such as the US more than doubled. In Australia some of the major improvements in Australian wheat yields have come since the 1960s, primarily attributed to the introduction of semi-dwarf lines from the international breeding programs, but also more advanced herbicide technology and lupin intercropping.

DDT was first synthesised during WWII in Europe as a 'safe, and durable' pesticide (Perkins 2008: 8) for malaria and lice control. After the war agricultural applications expanded, and a raft of closely related other pesticides, herbicides and fungicides were developed and released. Perkins (2008: 9) argues that these chemicals 'remade' human relationships with pests. Agricultural practices including crop rotations and water/fertiliser application rates changed. Labour relations changed as weeds could be sprayed instead of removed by hand. In retrospect we know of the multiple problems including toxicity, disease resistance and predator prey population problems. Hydrological development is less relevant to the Australian wheat story, but post war expansion of irrigation projects was particularly important for US wheat growing (Worster 1985, Reisner 1986).

Despite the advances in crop yield during this time, the 'promise' of higher yielding varieties to provide more food and wealth has been widely debated on social and political as well as environmental grounds. The scale and connection of plant breeding science and agricultural networks amongst industrialised nations has been critical in reinforcing multiple tiers of wealth generation and further investment in the science of plant reproduction and other agricultural technology (Perkins 1997). Perkins' equation of seeds plus nitrogen plus water was deceptively simple compared to the complex geopolitical edifice underpinning it. He argues that the push for high yielding wheat and cereal varieties was part of a larger geopolitical project, the containment of communism, marketed under humanitarian motivations to eliminate poverty and feed developing nations, from whom the most trenchant critiques have come (Shiva 1993). Because of its heavy reliance on international trade, Australia's wheat production is explicitly linked to these international stories of both agricultural commodity trading and hunger and food aid (Barrett and Maxwell 2006), as the example of Iraq in the next chapter also illustrates.

Post-war production surplus in fact presented something of a problem for the US, as international customers had little cash to buy their relatively expensive grain. It was problematic primarily because of the lack of infrastructure for, and technology requirements of, long term storage. To deal with inadequate storage capacity, some grain surpluses were fed back to livestock – recognised as an inefficient use of food, but something of an 'economic necessity' at the time.

Actually, feeding surplus grain to livestock began before WWII; it had been part of US agriculture since at least the 1880s. But it became significant in the 1930s when the practice caught the attention of grain company Cargill, who noted mid-western US farmers were having some success (such as faster growth rates) using surplus grain as stock feed by grinding it and mixing with added concentrates (Broehl 1992). Austen Cargill suggested this might be a way of dealing with periodic grain surplus and began innovating in stock feed formulas and production plants. As surplus and unsold grain became a more consistent post-war phenomenon, stock feeding with grain enabled meat to be produced more quickly and cheaply. This combination of factors contributed to the post-war culture of Americans eating more meat than ever before. It also contributed to a more concerted campaign of developing international markets for US wheat, in turn converting rice eaters in the developing world into bread and meat eaters.

The new Public Law 480 of 1954 was sold to US farmers and the American public on the basis that Americans would be providing global food aid, but it explicitly linked food aid to both farm and foreign policy. It facilitated a permanent international market for surplus US grain, allowing US farmers to be paid for continuously producing the surplus, rather than being paid not to produce it (Gilmore 1982). Under PL480 the US government 'authorised' foreign countries to purchase US grain with loans from US institutions. The loans had to be repaid but there were agreements that it could be with the borrowing country's own currency and over longer than usual time periods. Until the 1960s the US government was the principal financier of the grain trade, but Morgan (1980) argues that private trading companies gradually lobbied to change the legislation so that they could take over more of the deals the government was controlling. Morgan suggested this was promoted to the US public as the government freeing up trade and putting the US grain trade on a level playing field, thereby making agriculture and business more competitive. Gilmore (1982) argued that PL480 began the path towards a 'created' international market for wheat (and other grains), and in this process wheat became the internationally traded commodity as we now know it.

As an exemplar of this process, we can follow the story of Egyptian wheat, as explained by Mitchell (2002). (For examples from Russia, Nigeria and Korea see Gilmore 1982.) Mitchell takes issue with the conventional trope of Egyptian development as problematic, encapsulated as it is by a narrow strip of productive land along the Nile Valley and its Delta, occupied by a rapidly growing population, leading to projected food shortages. He shows that

> Between 1966 and 1988 the population of Egypt grew by 75 per cent. In the same period, the domestic production of grains increased by 77 per cent but total grain consumption increased by 148 per cent, or almost twice the rate of population increase. Egypt began to import large and ever increasing quantities of grain, becoming the world's third biggest importer after Japan and China. (215)

If agricultural production grew faster than the population, Mitchell asks, 'then why did the country have to import ever increasing amounts of food? The answer is to be found by looking at the kinds of food being eaten, and at who got to eat it' (Mitchell 2002: 213). He goes on to show that it was a 'switch to meat consumption, rather than the increase in population, that required the dramatic increase in imports of food, particularly grains' (215). Moreover, the increase in consumption of meat was not evenly distributed throughout the population but consumed mostly by tourists, non-Egyptian residents and middle to upper class urban residents (215). This shift was obscured in the official figures.

> Rather than importing animal feed directly, Egypt diverted domestic production from human to animal consumption ... Human supplies were made up with imports, largely of wheat for bread making. So it appeared as though the imports were required not to feed animals supplying the increased demand for meat, but because the people needed more bread. (Mitchell 2002: 15–16)

Mitchell shows in considerable detail how imported wheat from the US, depicted as aid for development, in fact supported US farmers, and how Egyptian 'government food policy forced even the smallest farmers to shift from self-provisioning to the production of animal products and to rely increasingly on subsidised imported flour for their staple diet' (217). Further, the depiction of land shortages as 'natural' allowed the question of land reform to be treated as insignificant: 'once the problems Egypt faces were defined as natural rather than political, questions of social inequality and powerlessness disappeared into the background' (221). Rather, 'Egypt's food problem was the result not of too many people occupying too little land, but of the power of a certain part of that population, supported by the prevailing domestic and international regime, to shift the country's resources from staple foods to more expensive items of consumption' (Mitchell 2002: 217).

Numerous authors have argued that post-WWII US agricultural and foreign policy paved the way for a system of agricultural and trade interdependence where food provision and self-sufficiency are no longer possible in many countries. Gilmore (1982) argued that eventual privatisation of this trade and concentration of power amongst the six major grain traders distorted the entire global grain market, in which supply is 'no longer matched with demand by efficient intermediaries' (Gilmore 1982: 4). Major companies and, to a lesser extent governments, distorted and manipulated the price of grain more than producers and consumers, and their success has been at the expense of both producers and consumers.

***

This chapter has focused on four important historical moments of wheat's becoming, exploring both continuities and thresholds of significant change. We have seen its shifting identity, both in its own genetic changes, and in human attempts to capture and categorise wheat. One of those continuities is that wheat was always a mobile plant, becoming international and global very early in its history. The characteristics of grassiness, or grassy plantiness, evolved to cope with climatic conditions, including drought and temperature extremes, which continue to characterise wheat landscapes. Breeding experiments today continue a history from tens of thousands of years ago, introducing alien and synthetic genes to produce even greater diversity in the wheat assemblage. But there was vernacular experimentation towards 'indigeneity' from the start, including Australian contributions back to the metropolitan centre. In the following chapter we examine how this history of becoming takes expression in the contemporary spaces of wheat.

Chapter 4

Spaces of Wheat

Beyond Goyder's line

In the spring of 1865 South Australian Surveyor General George Goyder was sent to examine the drought-stricken pastoral country to the north of Adelaide, with instructions to make such observations 'as may enable you to determine and lay down on a map, as nearly as practicable, the line of demarcation between that portion of the country where the rainfall has extended, and that where the drought prevails' (Meinig 1962: 45), in order that effective drought relief provisions could be administered. He returned a month later and produced the required map. The line coincided more or less with the southern boundary of the saltbush shrublands, extending considerably further south than he had anticipated. Donald Meinig's classic 1962 work *On the Margins of the Good Earth: The South Australian Wheat Frontier 1869–1884* uses archival analysis of parliamentary and media debates to analyse the northward expansion of the wheat frontier. He documented the way a line on the map – the famous Goyder's line – was empirically tested by colonising settlers during a period of climatic variability; expanding beyond it during wet years, then falling back again with the onset of drought.

Soon Goyder and others began to consider the line as having wider implications, in terms of separating the lands suitable for agriculture from those fit only for pastoral use. But in the early 1870s this was not yet a controversial decision; there was still at that time plenty of land available south of the line. Meinig wrote:

> As if to confirm the wisdom of making the lands more readily available, the rains came in good time, the hot winds withheld their searing touch, the red rust was rarely seen, and the crops of 1872–73 were superb. Sixteen to twenty bushels per acre were reaped all over the North and the Peninsula, and many localities surpassed this handsome average. A rash of complaints arose over the scarcity and cost of farm laborers, the shortage of rail trucks to the south, and the inadequate marketing facilities in the new areas. (Meinig 1962: 46–7)

In the following years there was something of a land rush, with expansionism supported by government policy and infrastructure investments. A series of good years increased the pressure, despite Goyder sounding dire warnings about unreliable rainfall. Newspaper editors and politicians were influenced by international interest in the idea that rain would follow the plough.

This agricultural expansion met with problems in the drought years of 1880–1882, when harvest failures and a great deal of hardship first stalled the advancing frontier, then sent it into retreat. Meinig summarised:

> a great geographical experiment had been made – an empirical testing of the qualities of the land, farm by farm, district by district. No longer need the arguments rage over where the limits of agriculture lay, for those limits, as well as the existence of the marginal lands, and the extent of the fertile, reasonably reliable country had been defined. Not that experimentation was at an end, but the general qualitative patterns of the agricultural region had been roughed out and further efforts would bring refinements rather than major alterations. (Meinig 1962: 206)

Kevin Dunn has noted the broader theme in Meinig's work of how European agricultural colonisation was frequently marked by a 'sudden influx ... followed by a straggling exodus, a quick testing of the land's productivity by ordinary people, often at great personal cost' (Dunn 2009: 48).

It is easy to read this story as a morality tale about the failure of European agricultural colonisers to read the land correctly in their haste to impose a northern hemisphere model of agriculture upon it. At one level that is unarguable, but as Meinig's work shows, the underlying story was much more complex. He argued that the South Australian farmer was in fact the pioneer of a new globalised agriculture. Their empirical experience of extracting productivity from the South Australian wheatlands constituted de facto resistance to the European yeoman ideal of a small freehold farm, 'worked by the family which it in turn supported from its own produce of field, garden, orchard, woodlot and livestock; yielding a modest surplus from a variety of crops carefully planted and tended' (Meinig 1962: 120).

In contrast, the South Australian farmer's wheat

> was not for his family and the village grist mill, it was wheat for the millions of the new industrial world. He farmed not as a member of an intimate, stable, localized society, but as a member of a world-wide dynamic, competitive society ... the invention of the stripper had opened the way toward this new agriculture, but a generation of experience had proved that in these precarious sub-humid lands under intensive competitive conditions it had to be grown on each farm by the hundreds instead of mere tens of acres. Labor, not land, was the scarce factor, and a whole array of new, enlarged, and efficient machinery now allowed a whole new scale of agriculture. (Meinig 1962: 121)

Meinig's approach was not without critique. A contemporary reviewer, Les Heathcote, noted that for all its strengths it was wanting in farm-scale detail; it contained 'the outlines of settlement history but not the body' (Heathcote 1963: 181). It is true that the voices of individual farming families are not heard against those of the editors and politicians. More striking to today's reader, given that Meinig went on to write famous works about indigenous Americans, is the complete silence on Aboriginal issues and people in the colonisation process in South Australia.

There has been much debate since about where the line really was, what it represented and whether it was in the right place. These issues are not the focus of our analysis. Rather, we use the example of Goyder's line to take issue with the question of lines – frontiers, boundaries, margins. What does it mean for us to analyse our agricultural systems, and the pressing socio-economic-ecological issues around them, mainly from this landscape view? And, following Heathcote's critique of Meinig, what might be hidden if other scales are obscured? In Chapter 3 we showed that the process of becoming wheat was always a spatial process. In this chapter we think more dynamically about the multiple spatialities of wheat, providing an overview of the global landscapes of wheat. The chapter also provides a transition to the Australian-focused material of the remainder of the book. We think about the spaces of wheat using three intersecting themes.

First, we draw on relational approaches to scale, as discussed for nearly two decades in geography (Howitt 1993, McGuirk 1997), to think through relationships between the local and the global. Previous understandings of scale held that it is 'a preordained hierarchical framework for ordering the world – local, regional, national and global' (Marston 2000: 220). These nested scales could be taken apart, or put back together; what Massey (2004: 9) referred to as 'Russian Doll' geography. 'The notion of nesting assumes or implies that the sum of all the small-scale parts produces the large-scale total' (Howitt 1993: 36). Instead, relational approaches understand scale as 'a contingent outcome of the tensions that exist between structural forces and the practices of human agents ... As geographers, then, our goal with respect to scale should be to understand how particular scales become constituted and transformed in response to social-spatial dynamics.' (Marston 2000: 220, 221).

Relationality challenges the idea that we can 'identify discrete scales from which causes originate and at which effects are felt. In such an approach processes, outcomes, and responses are categorised into distinct "boxes" that are seen as discrete entities originating at a particular level in an indisputable hierarchy of scales' (McGuirk 1997: 482). Thus the relationships between scale and order, or scale and causation, should not be assumed but be the subject of empirical enquiry. To say that scale is both socially produced and relational does not deny that particular scales can become fixed, reproduced, and influential – they can come to be seen as natural, as in the case of a globalised wheat market.

There are several implications here. We need to pay attention to the way in which different scales become constituted. For example, we illustrate different ways in which 'global' wheat flows are brought into being in the form of flows of data and maps. This process always takes place somewhere, through embodied practices of knowledge accumulation, translation and communication. We draw here on Mitchell's critique of the way 'economics takes for granted the nation-state as its object' (2002: 231) in his study of Egypt and modernity. Conversely, the 'local' does not just feed into pre-existing scales of something bigger, in accumulative fashion. Rather we also have the 'possibility of thinking of power as something local in construction; that is, drawing upon and shaped by larger logics, but built out of the practical relations between farmers and laborers, landowners and middlemen, bureaucrats and merchants, men and women' (Mitchell 2002: 167). Examining empirically the many different expressions of the 'local' as it applies to wheat, as we do in the remainder of the book, provides insights into the diversity of those 'practical relations', and the ways they constitute the larger logics of wheat.

In a second related theme, we consider what Plumwood called the 'shadow places' of wheat; 'the many unrecognised, shadow places that provide our material and ecological support, most of which, in a global market, are likely to elude our knowledge and responsibility' (Plumwood 2008: 139). The shadow places of wheat include many different landscapes of agricultural production, discussed in this chapter, as well as those places to and through which wheat moves to become food and other industrial substances, elaborated in Chapters 6, 7 and 8. In Plumwood's thinking, we must understand our places not just as those we 'love, admire or find nice to look at' (2008: 139), but also as the places 'that we don't have to know about but whose degradation we as commodity consumers are indirectly responsible for' (2008: 147). Later in this chapter we use the concept of ecological and water footprints to render visible a certain sort of wheaty shadow place; the environmental costs of its production and consumption.

Third, this discussion of spaces of wheat helps us position the Australian empirical material of the remaining chapters. It should be clear from the foregoing discussion that we are not attempting a panoptic overview of wheat that zooms down to an Australian 'case study'. Rather our perspective is that local, embedded experience and practice helps to constitute that which is then called national or global. As we illustrate, the fine grain of everyday life in 'remote' Australian wheat farms is often simultaneously local and global in perspective and connection. Inherent in our use of methods in the ethnographic tradition is an argument that 'starting in the middle of things' is one important approach to understanding the complexity of the whole.

None of this is to deny the importance of an international perspective or to downplay the global power of wheat. Rather, the breathtaking contemporary reach of wheaty landscapes demands that we pay serious empirical attention to how they have been brought into being, how they are stabilised and maintained, and how they might change in the future. In doing so, we continue to keep in mind the plantiness of wheat and its agency in these processes of spatialisation.

Global Wheat

Today wheat is grown throughout temperate and semi-arid areas of the world; indeed it is grown in more diverse environments than any other food (Mitchelle and Milke 2005). Wheat is grown over more land area than any other agricultural commodity (Table 4.1), its production area in 2010–2011 covering over 222.5 million hectares (USDA 2011). And since most of what is grown is *T. aestivum*, this is possibly the most extensive area occupied by any plant species on earth. Although global average wheat yields have increased significantly over the past 40 years, wheat has a lower average yield than either rice or corn, making its production more energy intensive per calorie produced.

Table 4.1 Global production, consumption and trade statistics for the leading agricultural food crops 2010–2011

	Wheat	Corn	Rice (milled)	Soybean
World Production area Million hectares	222.5	162.6	158.6	102.8
World Production Million metric tonnes	648.2	787.3	451.2	264
World Consumption (MMt)	655.3	843.3	446.1	252.9 (oilseed) 172.8 (meal) 40.96 (oil)
Total imports (MMt)	128.5	90.9	29.9	89.3 (oilseed) 57.7 (meal) 9.1 (oil)
Total exports (MMt)	131.3	92.0	32.7	91.2 (oilseed) 60.2 (meal) 9.9 (oil)

Source: USDA 2011.

As seen in Table 4.2, the world's wheat production is distributed across 10 or so main countries. For at least the last decade China and India have been the world's largest producers, growing 17.4 per cent and 11.9 per cent of global wheat respectively in the period 1996–2005 (Mekonnen and Hoekstra 2010: 1263), followed by the USA and the Russian Federation, France and Germany. This represents a significant shift since the 1970s and 1980s, when the USA and the then USSR were the world's largest producers, and from the 1930s when the USA overtook Australia as the largest producer (Broehl 1992). There are year to year differences in the order of the top ten producing countries (Table 4.2); for example Australia regularly drops out during drought years. This relatively widespread production distribution is different from corn, whose production is more spatially concentrated within the USA, and rice, whose production is heavily concentrated in China, India and south-east Asia (FAOSTATS 2011).

Table 4.2 Rank table of world's top ten wheat producing countries (tonnes) 1961–2009

1961	1970	1980	1990	2000	2001	2002	2003	2004	2005	2006	2007	2008	2009
USSR	USSR	USSR	USSR	China	China	China	China	China	China	China	China	China	China
USA	USA	USA	China	India	India	India	India	India	India	India	India	India	India
China	China	China	USA	USA	USA	RF	USA	USA	USA	USA	USA	USA	RF
India	India	India	India	France	RF	USA	RF	RF	RF	RF	RF	RF	USA
France	France	France	France	RF	France	France	France	France	France	France	France	France	France
Italy	Turkey	Canada	Canada	Canada	Australia	Germany	Australia	Germany	Canada	Canada	Pakistan	Canada	Canada
Canada	Italy	Turkey	Turkey	Australia	Germany	Ukraine	Canada	Canada	Australia	Germany	Germany	Germany	Germany
Turkey	Canada	Germany	Germany	Germany	Ukraine	Turkey	Germany	Australia	Germany	Pakistan	Canada	Ukraine	Pakistan
Australia	Australia	Pakistan	Australia	Pakistan	Canada	Pakistan	Pakistan	Turkey	Pakistan	Turkey	Turkey	Australia	Australia
Argentina	Germany	Australia	Pakistan	Turkey	Pakistan	UK	Turkey	Pakistan	Turkey	UK	Argentina	Pakistan	Ukraine

Note: USSR: Union of Soviet Socialist Republic; RF: Russian Federation; USA: United States of America; UK: United Kingdom.
Source: FAOSTATS 2011.

As a general pattern of export and import, wheat flows from developed agricultural producers (USA, Canada, France and Australia) (Table 4.3) to populous developing nations such as Egypt, Algeria, Indonesia and Brazil. But there are other important flows apparent in the data, for example significant volumes of wheat are imported by Japan, which is heavily reliant on imported food, and Italy, the largest wheat processing centre for pasta production. Additionally, large producers of wheat have sometimes been reliant on significant imports, for example China, and the USSR.

Table 4.3 Rank table of world's top ten wheat import and export (tonnes) countries 1961–2008

Rank top ten exporting countries					
1961	**1970**	**1980**	**1990**	**2000**	**2008**
USA	USA	USA	USA	USA	USA
Canada	Canada	Canada	Canada	Canada	Canada
USSR	Australia	Australia	France	Australia	France
Australia	USSR	France	Australia	France	Australia
Argentina	France	Argentina	Argentina	Argentina	RF
France	Argentina	USSR	UK	Germany	Argentina
Romania	Germany	UK	Netherlands	Kazakhstan	Germany
Sweden	Italy	Bel-Lux	Germany	UK	Ukraine
Ireland	Hungary	Hungary	Denmark	Turkey	Kazakhstan
Germany	Netherlands	Germany	Saudi Arabia	Belgium	UK

Rank top ten importing countries					
1961	**1970**	**1980**	**1990**	**2000**	**2008**
China	China	USSR	USSR	Brazil	Japan
UK	UK	China	China	Italy	Algeria
Germany	Japan	Japan	Japan	Iran	Egypt
India	Germany	Brazil	Egypt	Japan	Italy
Japan	India	Egypt	Italy	Algeria	Indonesia
Italy	Brazil	Poland	Iran	Egypt	Brazil
Brazil	USSR	Italy	Netherlands	Indonesia	Iran
Poland	Netherlands	UK	Algeria	Belgium	Morocco
Czech	Bel-Lux	Iraq	R. of Korea	Morocco	Turkey
Pakistan	R. Korea	R. Korea	Turkey	R. Korea	Spain

Note: RF: Russian Federation; USSR: Union of Soviet Socialist Republic ; USA: United States of America; UK: United Kingdom; Bel-Lux: Belgium Luxembourg; R. Korea: Republic of Korea.
Source: FAOSTATS 2011.

Most individual nations collect their own data, which are then assembled by the Food and Agriculture Organisation (FAO) of the United Nations, and also by the Foreign Agricultural Service of the United States Department of Agriculture. The FAOSTAT database boasts 'an on-line multilingual database currently containing over 1 million time-series records from over 210 countries and territories covering statistics on agriculture, nutrition, fisheries, forestry, food aid, land use and population' (FAOSTATS 2011). Our cut through this data is necessarily coarse. Representing the geography of detailed trade flows of wheat on the two dimensional page is not an easy task; see for example the FAO's interactive geographic mapping project for a detailed picture of agricultural commodity trade flows over the past 25 years (www.faostat.fao.org). The immense spatial area devoted to wheat production represented in these global statistics and maps should, however, be interpreted with care. They are not static spaces locked away from other kinds of production. Clearly the spaces of production and consumption are changing through decadal and longer temporal scales, reflecting a range of climatic, economic and geopolitical influences.

Australia

Australian wheat production is a significant component of this global network. In 2003–2004[1] Australia was the world's third largest wheat exporter, contributing 15 per cent of the global export market, behind the United States and Canada (ABS 2006). This was not the case with the 2006–2007 and 2007–2008 harvests, when severe drought led to significant reductions in export volume. The most significant recipients of exported Australian wheat by volume in 2009–2010 were Indonesia, Japan, Korea, Malaysia, China, Malaysia, Yemen, Iraq and Egypt (ABARES 2010) (Table 4.4); although the full picture of export distribution is even broader than this. Australia also imports wheat, mostly in the form of processed wheaten products from Italy, India, the UK, Turkey and the Netherlands (FAO trade flows wheat import maps 2011).

Production in Australia today surpasses all historical highs, but also continues to be characterised by years of significantly reduced productivity. In 2003–2004 the gross value of wheat production was estimated at around A\$5.6 billion (about 26 million tonnes (ABARE 2006)), representing about 15 per cent of Australia's total agricultural production (ABS 2006). However, production has been as low as 13 million tonnes (2007–08, ABARE 2009). In 2009, wheat was the country's third most valuable agricultural commodity behind beef and fresh milk, and is consistently the largest and most valuable crop produced. Despite this general increase in productivity, wheat is no longer a major contributor to gross domestic product; indeed agriculture in total represents only 2.5 per cent of GDP (ABS 2010).

---

1   The statistical reporting year in Australian agriculture reports the wheat grown throughout the previous winter and spring, and harvested in summer.

Table 4.4 Ranked Australian wheat and flour export destinations 1999–2000 to 2009–2010

1999–00	2000–01	2001–02	2002–03	2003–04	2004–05	2005–06	2006–07	2007–08	2008–09	2009–10
Iraq	Iran	Iran	Indonesia	Indonesia	Indonesia	Indonesia	Indonesia	Indonesia	Indonesia	Indonesia
Indonesia	Iraq	Iraq	Japan	Egypt	China	Egypt	India	Japan	Iran	Japan
Iran	Indonesia	Indonesia	Iran	Japan	Iraq	Japan	Japan	R. Korea	Malaysia	R. Korea
Japan	Japan	Japan	Iraq	Iraq	R. Korea	R. Korea	R. Korea	Malaysia	Japan	Malaysia
Pakistan	Korea	Egypt	R. Korea	R. Korea	Japan	Malaysia	Malaysia	Yemen	Yemen	China
Korea	Egypt	R. Korea	Egypt	China	Egypt	Iraq	Yemen	Egypt	R. Korea	Yemen
Egypt	Malaysia	Malaysia	Yemen	Malaysia	Pakistan	Thailand	Iraq	New Zealand	Iraq	Iraq
Malaysia	Yemen	Yemen	Malaysia	Thailand	Thailand	UAE	Egypt	Thailand	Egypt	Egypt
Yemen	UAE	New Zealand	New Zealand	New Zealand	New Zealand	Yemen	New Zealand	Kuwait	Bangladesh	Bangladesh
UAE	Thailand	Thailand	Thailand	Kuwait	Yemen	Kuwait	Kuwait	Iraq	Thailand	Kuwait

Note: UAE: United Arab Emirates; R. Korea: Republic of Korea.
Source: ABARE 2009, ABARES 2010.

Past studies of Australian wheat yields emphasised two features; declining yields in the latter half of the nineteenth century, followed by rising yields in at least two distinct periods since that time (see for example Donald 1963, Warren 1969 and Macindoe 1975) (Figure 4.1). Echoing the empirical testing of Goyder's line in South Australia in the late nineteenth century, both Western Australia and NSW saw phases of agricultural expansion into and retreat from more arid areas in the 1950s (Hamblin and Kyneur 1993). Within the national trend of yield increase after the Second World War there are periods of declining yield (Figure 4.1). There is also spatial variation in these gains not visible on our Figure; for example Hamblin's and Kyneur's (1993) analysis for 1950–1990 showed substantial differences both between and within states. Yield increases have been attributed to technological and soil management improvements (Warren 1969, Hamblin and Kyneur 1993, Robinson et al. 1999), the impact of wheat varieties and breeding (Warren 1969, O'Brien 1982, Antony and Brennan 1988) and other crop management issues (Kirkegaard et al. 1994).

Despite technological, breeding and other crop management advances (Hamblin and Kyneur 1993), significant variability remains a key feature of wheat yield in Australia. Yield is primarily driven by rainfall (Potgieter et al. 2002), itself enormously variable in Australia, as well as spatial differences in soil moisture potential (Stephens and Lyons 1998). Australian wheat yields are lower than in many other countries producing wheat in similar latitudes (Hamblin and Kyneur 1993, Calderini and Slafer 1998) and Australian wheat farms remain comparatively large in order to be viable. ABARE (2010) records an average farm size for 'grain producing farms' in 2008–09 as 2500 ha, although some farms where we interviewed in the southern district of our study area were over 11,000 ha.

To understand both change and stability in the export context of Australian wheat, it is necessary also to consider the background of single desk marketing, as explained by political scholar Linda Botterill. The Australian Wheat Board was the sole marketer of Australian wheat for several decades after it was established in 1948 (Botterill 2011).

> The Board's operations were based on ideas that can broadly be described as agrarian; belief in the essential worth of agricultural activity, suspicion of the middle man, and a basic egalitarianism that did not differentiate between grades of wheat or the skills of growers, and which valued the pooling of risk and equality of returns on a largely undifferentiated product. (Botterill 2011: 632)

The domestic grains industry was deregulated in 1989 'against the wishes of the majority of grain growers, removing the compulsory acquisition by the statutory Australian Wheat Board of all milling wheat grown in Australia' (Botterill 2007: 5).

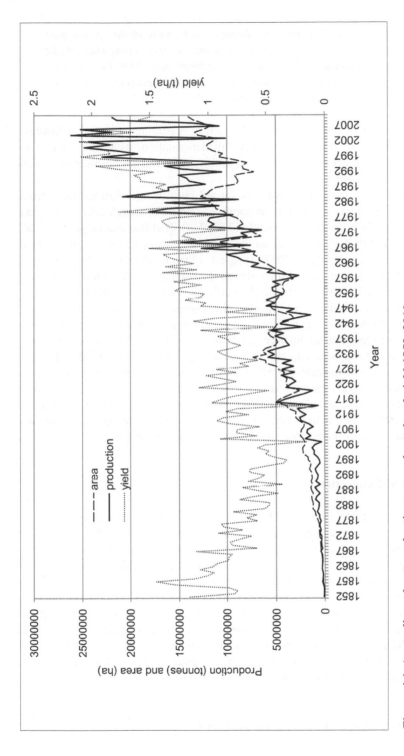

Figure 4.1 Australian wheat production, area planted and yield 1852–2010

Source: from Coghlan 1901, ABS (n.d.), Davidson 1987, Mitchell 2003, ABS 2006, ABS 2010. Historical data was standardised to metric units; decision rules for data sources were based on precision, accuracy and source authority.

This meant that the Board now had to compete in the domestic market to acquire wheat for export, where previously it was the only buyer of milling wheat. While the government removed the powers of compulsory acquisition, the Australian Wheat Board's powers as the single seller of export wheat (the 'single desk') were not altered [at that time]. (Botterill 2007: 5)

After 1989 the peak body, the Grains Council of Australia, undertook a strategic planning process to prepare for likely full deregulation, ostensibly aiming to deliver two key objectives; the retention of the single desk and grower control of wheat export marketing.

'Growers were concerned that a privatised body would shift its focus from maximising returns to growers to maximising returns to shareholders, who may not themselves be grain growers' (Botterill 2007: 6). That fear was largely realised. In 2001 AWB Limited was listed on the Australian Stock Exchange, attracting a broad shareholding. Botterill (2007: 6) reported that, by the end of 2006, growers owned less than 50 per cent of the value of the company, and argued that grower control was largely illusory (see also Botterill and McNaughton 2008).

Deregulation of bulk wheat exports – the abolition of the 'single desk' – took place in July 2008, increasing the options available to individual farming households as to when, where and to whom they can or should sell their wheat. As we see below and in Chapter 5, farmers continue to have a range of views about these changes, depending mainly on the extent to which they are comfortable in the role of business people as well as grain growers. It is easy in these overviews and statistics to lose sight of the conditions in individual paddocks and households, and their connections to global circuits of climate and capital. The moments of connection below provide insights into some of the complexities of these scales of analysis.

Moments of Connection

Fred has been working for the New South Wales Department of Primary Industries for as long as anyone can remember, advising farmers on the latest research about wheat varieties, and collating records of their experiences with his previous recommendations. Late in October 2007 he stands in a field of wheat in central NSW contemplating the prospect of another poor harvest. His mobile phone rings. It is a young woman from Bloomberg sitting at a desk in Melbourne, wanting an opinion from Fred on the state of the harvest. 'We're in pretty dire straits', he tells her; 'there's nothing out there to harvest, that's the reality of it'. As Fred tells the story, the next day the price of wheat futures in Chicago rose 3 per cent. As Bloomberg records it,[2] other factors were also involved, including the declining value of the US dollar, drought in Canada as well as Australia, and excessive rain in the US, France and Germany.

---

2   Bloomberg_com Oct 18 wheatrisesNews.htm.

***

Not far from Goyder's Line, a local company called Flinders Ranges Premium
Grain has spent much of the last decade growing the kukri wheat variety for the
frozen bread dough market. Dissatisfied with the prices it was getting for its wheat,
it pre-empted the 2008 deregulation of Australian wheat marketing by trying to
find a niche in the international market.

> The markets for frozen dough are worldwide – Japan, North America, Europe,
> Australia. Some of the big companies like Subway use frozen dough as the basis
> for their bread products. Many other companies use basically frozen-dough
> bakery products because it's a method of really retarding the development of
> the yeast in the dough and then, at a convenient time, baking it out to the kind of
> product they might like …

> In India's case, they were trying to use local flour and they were putting all sorts
> of additives in to try to make these quality frozen products. They ended up with
> a product that looked OK but it was like a rubber band to chew it. It just had
> no taste and it was very poor quality and we knew basically what their problem
> was when they explained it. It was the quality of the wheat they were trying to
> manufacture the flour out of. (Peter Barrie, the farmer who started the company,
> quoted in ABC 2009)

Dev Lall, who runs Baker's Circle, a modern manufacturing bakery supplying
Indian Subway chains, says he had tried the local Indian flour without success.
'Flour in India is milled for flat bread, leavened (sic) bread. It's perfect for that
kind of product. I don't think it was ever intended to be used to produce the kind
of breads and croissants to those that we are' (ABC 2009). He now buys hundreds
of tonnes of flour direct from the Flinders Ranges farmers.

While Australian wheat was being rolled out in Indian Subway shops, at
the time of that interview, customers in Australian Subway shops were eating
Canadian wheat processed into frozen dough in the small New Zealand township
of Manaia. 'The superior qualities of Canadian grain means the extensibility of a
product allows for optimum product volume and very long shelf life, frozen. So
we've never considered using Australian grain because it could never match those
characteristics' (Paul Yarrow, New Zealand baker, interviewed by ABC 2009).

Mr Yarrow had commenced baking trials with the Flinders Ranges Kukri
wheat. If this was successful, it would potentially increase the market share for
Flinders Ranges Premium Grain exponentially.

***

As we saw above (Table 4.3), Iraq has been one of the most significant recent
destinations for Australian export wheat. However the mechanisms by which

this happened were only brought to light by a drama that played out far from the nation's wheat paddocks from December 2005 through to the end of 2006. The Cole Inquiry, officially known as the 'Inquiry into certain Australian companies in relation to the UN Oil-for-Food Programme', began taking evidence, handing down its report on 24 November 2006 (Overington 2007). The inquiry was into events as far back as 1999, when Australian Wheat Board (AWB) officials began paying kickbacks to the regime of Saddam Hussein in the form of 'trucking fees' attached to wheat contracts. Under the 'Oil-for-Food' program, Iraq – otherwise trapped by United Nations economic sanctions, to which Australia was a signatory – could sell oil, provided the money was used only to buy food and medicine, as monitored via a UN escrow account. But Iraq had requested, and AWB officials had agreed, that an extra $US12/tonne (later up to $US45/tonne) be paid in cash, on a 600,000 tonne shipment of wheat. 'In 1999–2000, for example, AWB had sold a record 2.4 million tonnes of wheat to Iraq. AWB had, in effect, secured 98 per cent of Iraq's wheat market' (Overington 2007: 54).

Why would Iraq want 98 per cent Australian wheat? No doubt it had much to do with the single-minded and ferocious techniques of the AWB grain salesmen, who often referred to the imperative of maximising grower returns. 'Many witnesses defend[ed] their actions in terms of the necessity of delivering results to Australia's grain growers by ensuring continuity of business in a very important export market' (Botterill 2007: 5). In doing so, they seem not to have reflected on the wider context of their actions; both the grain salesmen and the responsible politicians remained unapologetic throughout the process. Botterill (2011: 637) argues that the culture of the organisation changed after the 1989 domestic deregulation from being 'anchored in collective agrarian values', to being 'prepared to sell wheat whatever it took'.

But it was also to do with the qualities of the wheat itself. According to Overington, the long-time director of the Iraqi Grains Board, Zuhair Daoud, the central Iraqi player in the original negotiations, had over the years 'developed a preference for the high-quality, hard, dry Australian version of the grain, so much so that the Iraqi palate had actually adapted to the texture and flavour of Australian wheat' (Overington 2007: 7). This wheat would be milled together with domestic wheat to go as flour into the monthly ration baskets on which 80 per cent of Iraqis depended for their survival.

Inside the Wheat Belt

Australian wheat is grown in every state and in the Australian Capital Territory, across some 13.9 million ha in 2009–2010 (ABS 2011). This area is collectively referred to as the wheat belt (also the wheat-sheep belt) (Figure 4.2). The 'belt' extends in two crescents through south-eastern and south-western Australia, in areas with annual average rainfall between 400 and 600 mm, most of it falling in winter and spring. In the southeast this crescent is on the western (inland) side of the Great Dividing Range, extending from central Queensland through New South Wales (NSW) and Victoria, and into South Australia. It is important to note that in our detailed study area – the NSW part of the wheat belt – many farmers have more domestic marketing options than those in the more sparsely settled Western Australia and South Australia. In eastern and central parts of the NSW wheat belt, proximity to large regional towns means that there is a market for wheat for stock feed, milling and pet food manufacture. (Options decrease further inland as the cost of road transport becomes prohibitive.)

Production in the wheat belt was built not only on the removal of Aboriginal owners, as discussed in Chapter 3, but also on the broad-scale clearing of native vegetation and its associated environmental impacts, including loss of biodiversity, salinity and soil erosion (Young 2000) (Figure 4.3). Wheat landscapes are built upon the ecosystem most often represented as the quintessential Australian bush; an extensive band of grassy woodlands typically dominated by eucalypts over a diverse perennial grass understorey (Keith 2004). Main (2005: 65) cites a 1909 newspaper report describing heavily timbered land becoming 'hundreds of acres of grand wheat land without ... the sign of a tree or a stump upon it'. Grassy woodlands are characterised ecologically by episodic regeneration in response to periodic bush fire and dynamic cycles of drought and flooding (known as disturbance-gap dynamics), as well as other influences such as grazing pressure (Keith 2004). This diverse assemblage of vegetation communities historically supported large populations of arboreal mammals, such as koalas in NSW, as well as immense populations of nectar-feeding and seed-eating birds (Keith 2004). Between 60 per cent and 90 per cent of these woodlands have been cleared since European settlement. In response to the impacts of broadscale clearing (fragmentation, loss of cover and structural complexity, changes to fire regimes), many of the species previously present in the wheat belt have either become extinct, or are characterised as populations under threat; grassy woodlands are commonly described as 'the continent's most damaged and threatened ecosystems' (Keith 2004: 84).

Consequently it is common today to find wheat and grassy woodlands occupying the same space, but on completely different maps. With the exception of Ellis and Ramankutty's biome reconceptualisation, discussed in Chapter 2, crops are rarely depicted on maps of Australian vegetation. The conceptual divides between native and non-native, and before and after European settlement, are emplaced starkly on Keith's (2004) native vegetation maps for NSW, on which the spaces of wheat are represented as white or blank. So, the most grassy of grasses is not considered to belong. Nor do we find eucalypts or native grasses on maps of the wheat belt, which is usually conceptualised in terms of commodities that flow from it. This converse conceptual divide overlooks the productive contribution of these ecosystem services: 'native vegetation plays a fundamental role in binding soil and cycling nutrients... as well as controlling salinity, pests and diseases ... functions which indirectly and directly influence economic viability and sustainability of Australian farms' (Keith 2004: 14).

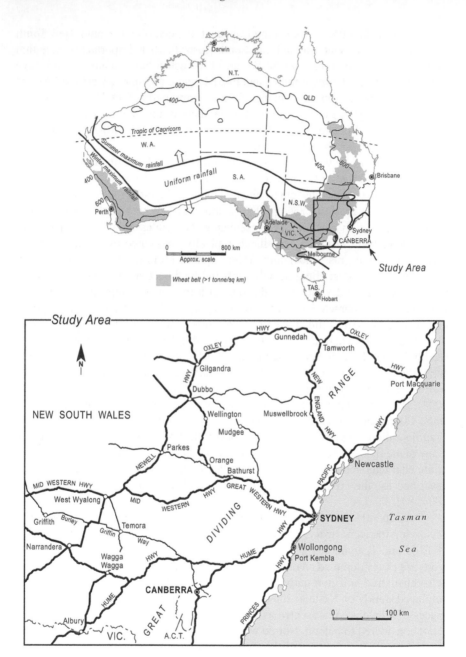

Figure 4.2 Location of the Australian wheat belt, rainfall zones, and study area
Source: Adapted from ABS 2006: figure a, BOM 2005: Australian average rainfall, annual
and Jeans 1987: summer winter rainfall zones.

Figure 4.3 Clearing timber on 'Sherwood' Ariah Park, NSW, c1920
Source: Mitchell Library, State Library of NSW, bcp_03096.

These white spaces of emptiness on both types of map are illustrative of a separationist paradigm between Australian agriculture and environmental protection (Saltzman et al. 2011). As Cocklin and Dibden (2009: 5) have argued, 'there has been 'a general tendency in Australia to under-value biodiversity on *farmed* (contrasted with *wild*) lands' (emphasis in original). The separationist paradigm ignores important on-ground assemblages that combine critically endangered ecosystems with spaces of economic production. Although many have found them too messy or difficult to think about, there is increasing recognition of ecological and landscape interconnectivity within the ecological sciences (see for example Manning et al. (2009) on landscape fluidity, Lindenmayer et al. (2008) on novel ecosystems, Fischer et al. (2010) and Gibbons et al. (2008) on the value of paddock trees). Such assemblages are at the 'frontline' of major efforts at landscape restoration, revegetation and environmental remediation that simultaneously challenge both agricultural and conservation policy.

It is now acknowledged that biodiversity conservation objectives cannot be fulfilled on public lands alone. As 70 per cent of land in Australia is privately owned or managed, and there is agricultural activity on 54 per cent of all land, the attitudes and practices of farmers and graziers are critical to issues of biodiversity and environmental management (ABS 2010). A range of innovative partnerships that connect habitat in fragmented landscapes will be needed, along the lines of Indigenous Protected Areas and Voluntary Conservation Agreements (Adams 2008). A more systematic acknowledgement of multifunctional agricultural landscapes will become necessary in order to support and scale up the efforts of farmers who are protecting or restoring habitat on their land, as reported in the agricultural censuses of 2005–06 and 2006–07 (ABS 2010). For example, the agricultural census of 2005–06 reported that 4.2 million trees were planted by farmers across the country for nature conservation, and a further 6.3 million for 'the protection of land and water areas' (ABS 2010: 507). In 2006–07 65.8 per cent of farmers reported undertaking improvements to their natural resource management practices (ABS 2010). Fifty thousand land managers across the country reported setting aside 9.2 million hectares – an area two-thirds the size of the wheat belt – specifically for conservation and biodiversity protection (ABS 2010). It is known that these activities are expensive and can reduce farm profitability (House et al. 2008). Although the details of these 'improved practices' need to be further researched, and the 9.2 million hectares must include a lot of grazing rather than cropping land, the scale of this potential conservation enterprise has yet to pervade the general Australian consciousness. That much of this activity coincided with drought years is particularly interesting. On a purely practical level, drought years give farmers time to do other kinds of work. More broadly it reflects the interactive agency of drought and farmers in accommodating workable long term ecologies for the land.

The Footprint of Wheat

The concept of the ecological footprint (Wackernagel and Rees 1996) was developed to bring to light the invisible or undervalued ecological underpinnings of contemporary lifestyles and products. The footprint provides an estimate of the area of productive land required to produce a product, or to sustain a person, a city or a nation. A range of tools are in use, all of them undergoing continual refinement due to the inherent difficulties of measurement. Environmental Life Cycle Assessment (LCA) is a related methodology that produces findings against a range of environmental criteria, rather than the summary number of the ecological footprint.

Narayanaswamy et al. (2004) undertook an LCA for the wheat to bread pathway in Western Australia, using a cradle to grave approach. Environmental impacts were assessed along this pathway in four phases (2004: 5):

- Pre-farm production of farm supplies and grain production (pesticides and fertiliser production, wheat growing)
- Storage, handling and processing (grain storage, flour milling, baking, includes packaging)
- Retail and consumption, including retail operations, storage at retailer and consumer, and consumption, including discarding the packaging
- Transportation (fertiliser and chemical transport to farm, grain transport to storage and processing, transport of grain intermediates, transport of packaged grain products to retailers and end-consumers)

Six different types of environmental impact were assessed; Resource Energy, Global Warming, Atmospheric Acidification, Human Toxicity, Terrestrial Ecotoxicity and Eutrophication (Narayanaswamy et al. 2004: 22). This nuanced depiction of some of the 'shadow places' of bread production helps us resist simplistic approaches to more sustainable food production, and start to think through how the responsibility for dealing with impacts should be shared across the community. The pathways varied in their proportional impacts:

> Retail and consumption was the most predominant subsystem in Resource Energy (~54%), Global Warming (~56%), Atmospheric Acidification (~77%), and in Human Toxicity (~40%) impacts. Pre-farm and farming was the most predominant subsystem in Eutrophication and Terrestrial Ecotoxicity impacts (~95% each). Storage and processing contributed 38% to total resource energy, 27% to total Global Warming, 20% to total Human Toxicity and 17% to total Atmospheric Acidification. Transportation contributed 10% to total Human Toxicity and its contribution to other impact categories was insignificant. (Narayanaswamy et al. 2004: 22)

Identification of hot spots (>10 per cent of contribution to the given impact) allows effective mitigative action to be targeted. For example, hotspots for bread are electricity use in retail and consumption, and in storage and processing, and insecticide use and fertiliser use in crop production. The low proportion of impacts associated with transportation challenges the ideal of local food as a solution (see also Penker 2006: 377, Saunders et al. 2006, Roggeveen 2010). Evidence in a study from production to port indicates that the most significant contribution to greenhouse gas emissions (35 per cent) comes from fertiliser production in the prefarm stage, followed by on-farm CO_2 emissions (27 per cent) (Biswas et al. 2008). Certainly changes to on-farm practices could reduce these detrimental effects, as well as the damaging impacts on terrestrial and aquatic ecosystems of fertiliser and pesticide use. But these appear to contribute much less to the Global Warming category than do the storage and processing phases of flour milling, and the retail and consumption phases.

A different take is presented by the water footprint analysis of Mekonnen and Hoekstra (2010), where *blue* water is the volume of surface and groundwater consumed (evaporated), *green* is rainwater and *grey* is 'the volume of freshwater required to assimilate the load of pollutants based on existing ambient water quality standards' (2010: 1,259). This analysis provides a way to trace the flows of virtual water in the world wheat trade, again with results that challenge the idea that localised production is necessarily better. Mekonnen and Hoekstra argue that the proportion of world wheat being produced using green water (currently 70 per cent) should be increased where possible, and the proportion of blue water reduced.

> Green water generally has a low opportunity cost compared to blue water. There are many river basins in the world where blue water consumption contributes

> to severe water scarcity and associated environmental problems, like in the Indus and Ganges basins ... since wheat has relatively low economic water productivity ... one may question to which extent water should be allocated to wheat production in relatively water-scarce basins. (Mekonnen and Hoekstra 2010: 1,264)

As wheat exports around the world are predominantly from rain-fed agriculture, international trade can be considered to be contributing to global average water savings. 'Import of wheat and wheat products by Algeria, Iran, Morocco and Venezuela from Canada, France, the USA and Australia resulted in the largest global water savings' (Mekonnen and Hoekstra 2010: 1,265). Or, put another way, 'without trade the global wheat-related water footprint would be 6 per cent higher than under current conditions' (2010: 1,273). Whether it is considered good or bad to be a virtual importer or exporter of water in relation to wheat depends on the rainfall characteristics of each country. The trade in durum wheat from France to Morocco is presented as another example of a trade that contributes to overall water savings.

Conversely, a high water footprint of wheat consumption per capita can be explained by high consumption (for example Australia) or the wheat consumed can have a high water footprint (Kazakhstan), or a combination. There are complex implications here, not to mention difficulties of data measurement. Mekonnen and Hoekstra observe that 'the costs of water consumption and pollution are not yet properly factored into the price of traded wheat, so that export countries bear the cost related to wheat consumption in the importing countries' (2010: 1,273). They cite Japan and Italy as examples of countries with high external water footprints which 'put pressure on the water resources of their trading partners' (2010: 1,273). Japan imports 93 per cent of its water footprint, mainly from the USA (59 per cent), Australia (22 per cent) and Canada (19 per cent), while Italy imports virtual water from places that include water-scarce Middle Eastern countries such as Syria and Iraq (2010: 1, 270–1).

Of course, water is only one of the impacts associated with our production and consumption of wheat. Like greenhouse gas emissions, water accounting is likely to become more important over the coming years. This may mean we have to juggle competing requirements in decision-making. But at the moment none of these are factored into pricing or decisions, which are purely on dollars and protein levels. The complexity of allocating responsibility in different types of footprint and life cycle analyses has led practitioners to use the concept of shared responsibility, encouraging both producers and consumers to enter into dialogue about how to make improvements together (Lenzen et al. 2007).

***

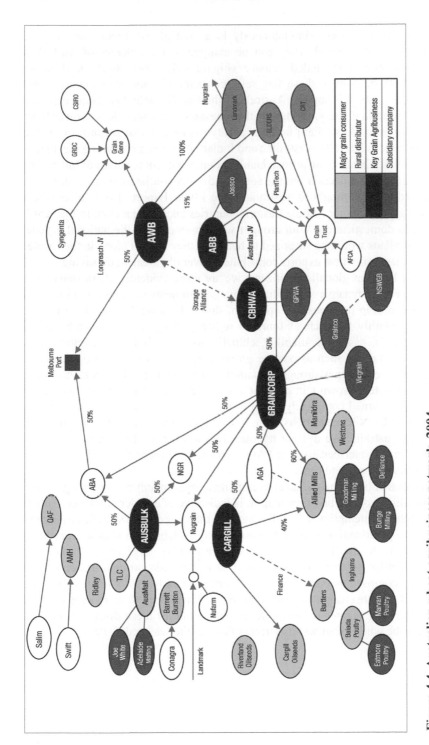

Figure 4.4 Australian wheat agribusiness network, 2004
Source: GRDC and GCA 2004: 6.

The spaces of wheat are simultaneously local and global, hidden and visible, tangible and metaphorical. They can be mapped as the linkages of Australian agribusiness and state-funded science (Figure 4.4). They swirl in daunting complexity through even these few selected examples, and over hundreds and thousands of years. It is not only Australian wheat and flour that are being exported to India; Australia is repaying some of the longstanding genetic debt it 'owes India when it comes to [Australian] farmers' ability to produce vast surpluses of bread wheat' (Braidotti 2011: 9). Australian and Indian scientists are working together to test Australian wheats in the unsustainable groundwater source areas of India. This circle of genetic connection returns some of the stress-tolerant genes that Indian wheat biodiversity brought to Australia in the 1890s. Export of Australian wheat to Iraq, Egypt and other Middle Eastern countries takes wheat back to the hearth areas of its domestication. With drought well entrenched in the national psyche, most Australians would be surprised to hear that they are not alone in their water stress, or that their wheat exports contribute to overall global water savings.

At one level, as globalised citizens, we are more evidently comfortable with these connected, dynamic spatialities. State governments no longer send men on horseback to map the lines and spaces of drought. In NSW, drought maps are produced monthly for each of fourteen regions, taking into account historical rainfall records, pasture availability, climatic events such as rainfall and frosts, and seasonal factors such as pasture growing seasons (http://www.dpi.nsw.gov. au/agriculture/emergency/drought/situation/drought-maps). Maps that had been almost completely brown ('in drought') or beige ('marginal') in the summer of 2006–07 had turned completely green ('satisfactory') by Christmas 2010, when the return of La Niña conditions brought cyclonic rainfalls and flooding to much of eastern Australia. We did not try ourselves to map the trade flows of wheat earlier in the chapter because the FAO offers readers more thorough interactive possibilities in space and time.

On the other hand, the spatial and temporal dynamism inherent in trying to even conceptualise drought, and then map it, is indicative of a wider challenge. This is the difficulty of attempting to stabilise agricultural systems and societies enough to keep them operational in a context of underlying variability in almost everything, or conversely to develop flexible systems responsive to change. Climatically speaking, settler Australians have been slow in this task; drought has only been formally recognised as part of normality since the early 1990s (Hennessy et al. 2008), and climate change suggests that 'normal' will need to be redefined again. In the following chapter we examine the interaction of these diverse processes, following the wheat plant and its farmers through the seasonal cycle.

Chapter 5
Growing Wheat

The New South Wales Wheat Belt, 2006–2008

Somewhere in the NSW wheat belt, Andy looks and talks like any other stoic and laconically good-humoured farmer dealing with the vagaries of the worst drought in living memory (Figure 5.1). But when Andy rises early, he turns on the espresso machine with one hand and the computer with the other. Before the coffee is brewed, he has checked the price of wheat on the Chicago Board of Trade along with the weather forecast, and a host of other domestic financial information. It is a morning ritual repeated with minor variations amongst many of the farmers of his generation. He will later drive the half hour or so from his home in a regional town, where his wife works, to the family farm where his elderly parents still live. In doing so Andy is juggling multiple temporalities – his daily and seasonal farm schedules, the aspirations of he and his wife for their life together, and the long term movements of capital.

Figure 5.1 Good-humoured, stoic Andy

Somewhere else, Peter and his wife Heather can see the wheat paddocks from the window of the kitchen where they share a noisy breakfast with their young children. Peter speaks of the great satisfaction in producing a good crop of wheat:

> ... you know, just looking out there and knowing the decisions you have made regarding fertiliser and spraying and when you have sown it, that all those things have come together to make a good crop. And there are plenty of things that will make it not a good crop. So making the right decisions and ending up with a good article at the end is very satisfying, yeah.

Later, over a cup of tea, Heather tells us of Peter's stress-induced headaches and sleeplessness over the previous few months, as a result of forward selling of a wheat crop that did not eventuate. We had interviewed them first in March 2007. By December when we visited again, late rain had delayed the harvest of what little wheat there was. In the intervening months 'every weather forecaster known to man was forecasting that the drought had finished and that this was going to be a higher rainfall year', and the household had forward sold part of their crop. With humour as dry as the paddocks around her had been, Heather said, 'They weren't wrong, they were just a bit late.'

As we talked to farmers across the wheat belt, it emerged that there is no single right way to grow a crop of wheat. Each farmer makes a series of strategic responses to a perceived set of conditions, some of which (for example, soil moisture and nitrogen levels) can be known, others (wheat prices, amount and timing of rain) of which have to be guessed. As Kevin, a private agronomist, says of his clients, 'Each person is different, some guys they're just business people, and there's some people who are lifestyle farmers and there's others who are very cost focussed. There's no blanket approach.'

What they have in common is the life of the wheat plant and its seasonal cycle of germination, growth, flowering and fruiting. Given the technological complexities of the agro-industrial edifice it seems somewhat quaint that the plant is still pinned to the soil, dependent for its plantiness on essential nutrients provided by rain and sunshine in the right quantities and at the right times. But it takes much more than this to produce a crop. Other plants keen to take advantage of the ploughed paddock must be killed, as must insects and pathogens that would treat the wheat as food. All of this takes money and machines, made available to the farming household via complex loans and other financial instruments, as well as the commodity prices they receive for the finished product. Farming households perform a complex balancing act between raising the capital to make necessary investments in the crop and their aspirations to self-reliance and low debt levels.

The growing of wheat is a complex coproduction between human, plant and other bodies, and the constellation of other things and processes that need to come together. In taking that coproduction as the theme of the chapter, we also emphasise multiple temporalities. Some of these are juggled by the farmer; from the decision about what to plant this year, and when, to the timing of loan repayments, to generational succession planning. Others are outside human control; the pulse of rainfall events, the snap frost, the rhythm of drought. Farmers have a variety of sensibilities to the inherent riskiness of trying to match human temporalities with the wider swirls of climatic processes. Kevin describes some of his clients, for example, as 'really switched on'.

> They're planning ahead because they either know that it's going to rain, they're more positive it's going to rain at some stage so they're ... getting ready to do that. And then there's the other people who will wait, and they're always behind the eight ball ... A lot of it's timing. That sometimes is the daily difference between a good farmer and a bad farmer ... being in on time. (Kevin, private agronomist)

Of course, the aggressive optimists can get caught out if the rain does not eventuate. We analyse these risk sensibilities at the end of the chapter, in preparation for discussing their relevance to the risks posed by climate change in Chapter 9.

Our study area encompasses southern New South Wales, the most productive part of the southeast wheat belt crescent, and part of northern NSW. Planting occurs in the southeast wheat belt after a hoped for 'autumn break' (rainfalls that come at the end of summer), then depends on winter and spring rainfall for growth, prior to an early summer harvest. See for example the 2006–2007 production statistics in Figure 5.2, which reflect the wheat harvested mainly in November-December of 2006. In the past this has proved reasonably successful with an approximate one year in seven drought cycle, when a poor harvest is often followed by a bumper harvest. Older farmers compared the 2006–2007 harvest failure with the drought of the 1940s, particularly the failed harvest of 1946. However, in the early 2000s a number of farmers have experienced only one good year (2005), leading to long term decline in soil moisture levels. In 2007, early good rains led to much speculation of just such a bumper harvest, but hopes evaporated as the winter and spring rains failed to materialise. For those who did manage to harvest some grain in 2007, the financial outcomes were not as poor as the year before, because in the intervening twelve months the price farmers could get for their wheat had risen considerably.

While drought is a regular feature of the Australian environment, droughts vary in their duration and extent. In Australia, exceptional circumstances (EC) are declared when:

> [R]are and severe events outside those a farmer could normally be expected to manage using responsible farm management strategies occur. Events must be rare, that is not have occurred more than once on average in every 20–25 years; must result in a rare and severe downturn in farm income over a prolonged period of time; cannot be planned for or managed as part of farmers' normal risk management strategies; and must be a discrete event that is not part of long-term structural adjustment processes or of normal fluctuations in commodity prices. (DAFF 2009)

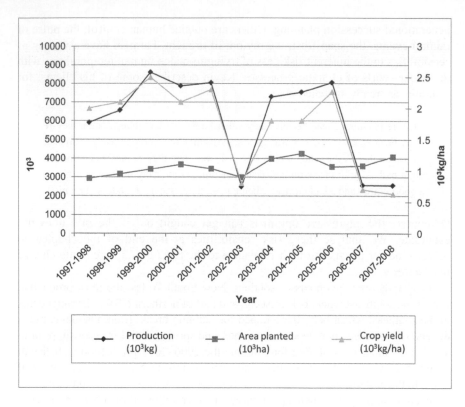

**Figure 5.2 Australian wheat production, area planted and yield 1997–1998
to 2007–2008**

Source: ABS 2008a and ABS 2008b. The statistical reporting year in Australian agriculture
reports the wheat grown throughout the previous winter and spring. Thus for example the
2007–2008 year reports the growing season through the winter of 2007.

In July 2008 the Federal Minister for Agriculture declared nearly two thirds of
Australian agricultural land to be covered by 74 EC declarations (MAFF 2009).
Our study years thus provide a research window onto climate change scenarios of
more frequent droughts.

Wheat farmers are responding to drought and climate change in a deregulated
market of 'competitive productivism', the outcome of an export-oriented
agricultural sector and a neoliberal political orientation (Dibden and Cocklin
2005, 2009, Andree et al. 2010). Ongoing deregulation of the sector culminated in
the 2008 abolition of the 'single desk' arrangements, under which the Australian
Wheat Board (AWB) had held the monopoly on export sales. Farmers in our study
area also have a number of options for domestic sales, including sales of wheat for
flour milling, stock feed processing and industrial manufacturing. These choices

and options bring with them a significant office workload that has an impact on everyday farm life.

We interviewed 29 farming households, from Albury in the south to Gunnedah in the north, and from Orange in the east to Griffith in the west (Figure 4.2, Figure 5.3) (Table 5.1). Participants were approached through personal contacts in the private agronomy industry, and also recommended by government agronomists in the New South Wales Department of Primary Industry (DPI). Our sample encompasses most of the socioeconomic diversity in the region. During fieldwork in December 2006, March 2007, May 2007, November-December 2007, and November-December 2008, 17 participants were interviewed once, 11 twice and one three times, depending on availability (a total of 42 interviews). Taped interviews lasting one to two hours usually took place in farmers' homes, and were often followed by a walk or drive around the farm. During the two December trips in the south (2006, 2007), the poor harvests gave farmers more time to talk to us. Durum growers in the north were interviewed in late November – early December 2008, in the middle of harvest season.

Figure 5.3 Interviewing on the back verandah

In 10 cases we were able to talk to couples who manage the farming enterprise together. On the whole however, farming was presented to us as a very gendered activity. With few exceptions, both men and women described the male/s of the household as the decision-makers and farmers, the women managing the household and childrearing. The exceptions were several more wealthy households – those that we later label strategic – where the women also managed the finances and the business side of the farming enterprise. We also had several examples where two or three generations of the family are working together, so we could discuss succession issues and elucidate generational differences in educational background and approach to risk (see for example Households K and T, U and V, W, Table 5.1) (Figure 5.4). In some examples the older generation has retired off the land and is now living in town. 'Dad' or 'granddad' had a presence in a number of interviews without being there physically; farmers talk about how 'he' did things differently or thought differently about things. On the other hand some of the study participants are not from the bush at all, but came from the city to be farmers (such as household Y).

Figure 5.4 Younger generation farmer

Table 5.1 Summary characteristics of farming households interviewed, with pseudonyms used in the text

Code	Pseudonym	Area	Interviewee/household structure
A	Don	NC	Male 50s/family with young son
B	Chris	C	Male 40s/family with young children
C		NC	Manager/Corporate farm
D	Jack	C	Couple 60s
E	Ted	C	Male 40s/also CEO of related businesses
F	Keith, Fiona	NW	Elderly couple/adult children off farm
G	Joseph	NC	Bachelor brothers 50s
H	Arthur	CW	Elderly couple
J	Terence	NC	Retired male
K		C	Elderly/father and father-in law of T
L		NW	Male/neighbour of N
M	Jim	S	Male 60s/farms with son and brother
N	Susie	NW	Couple /adult children employed off farm
O	Andy	C	Male 40s/working parents' farm, wife employed off farm
P		S	Elderly couple
Q		C	Elderly couple/son now farming their land
R		NW	Couple 40s/family with young children
S		C	Retired female
T	Vince	C	Daughter and son-in law of K, couple with young children
U	Gordon, Mark	C	Father and son
V	Patrick	C	Retired couple/Father/mother and grandfather/mother of U
W	Ian	S	Couple 50s and son/adult children on farm/related business
X	Janice	NC	Female 50s/adult children on farm/related business
Y	Peter, Heather	NW/NC	Couple 40s/family with young children
AA		FN	Couple 40s/family with young children
BB	Zack	FN	Couple 40s/family with young children
CC		FN	Father and Sons
DD		FN	Couple 40s/family with young children
EE	Dave	FN	Couple 50s with adult children employed off farm

Note: NC = North Central, C = Central, NW = North West, CW = Central West, S = South, FN = Far North.

One limitation of this research is that those farmers struggling most are also the most difficult with whom to make contact. Our initial points of connection – agronomists, DPI or farmer advisors – are themselves best connected with those who seek their expert advice and guidance, or who belong to farmer management groups who regularly share lessons from their successes and failures. There is something of a vicious circle whereby the least resilient farmers are least likely to have connections to this network; certainly they cannot afford private agronomic advice. Capturing the important views of those who were 'doing it tough' occurred through specific requests and social or neighbourly contacts of other participants, but these are probably under-represented in our sample.

Wheat and its Companions

Wheat in this region is always grown as part of a bigger farming system, whether a rotated cropping system or a mixture of cropping and livestock. Where stock are absent there is no requirement for fences, so paddock sizes may be much larger, or even non-existent. None of the durum growers we interviewed ran stock, considering them to interfere with cropping systems by degrading soil profiles. One grower explained that 'where crops are, stock don't go' due to the soil compaction and damage which results. In other cases stock are integrated, and the pre-flowering wheat crop is crash grazed to fatten lambs or cattle. For some farmers these decisions are lifestyle choices; stock require constant attention and are seen as too demanding, unlike wheat which gets on with its own life and does not need constant human attention.

> … we used to … have a lot of sheep and produce wool and I've gone into crops because you can actually get away … So I've got rid of all the ewes, which is why I got into trading, so I can have sheep when I want them and it suits me and then I can sell them and have absolutely none, and then we can just go away and do what we want to do. It's a lifestyle choice rather than economic. I'd probably make more money if I bred sheep but I live here [in a regional town], I don't live out there and it just gives you the freedom to go and do what you want to do. (Andy, household O)

In southern NSW canola (*Brassica* spp.) is commonly grown in conjunction with hard bread wheats and soft biscuit wheats, because canola and wheat are vulnerable to very different types of plant diseases. Further, the canola plant has a long tap root which rots away facilitating the growth and penetration of the wheat roots. Durum wheat cannot be grown over a previous durum crop due to its high susceptibility to crown rot, a fungal disease affecting plants in drier soil conditions and preserved on previous durum crop stubble. The crops most commonly grown in the northern durum area included bread wheat, sorghum and cotton, with mung beans and other specialty crops occasionally used.

Seasons

Although the broad cycle of wheat growth occurs in the winter and spring months, there is a lag of six to eight weeks along a transect from north to south. Northern NSW experiences longer winter sunlight hours, warmer conditions and higher rainfall. In addition the heavier clay soils of this region retain more moisture than the sandy loams of the south. This means that in the northernmost reaches of the wheat belt, the season is so compressed as to allow a summer crop to be grown between the harvest and sowing of wheat. In the sheep belt to the south, as soon as harvest finishes, shearing begins.

The cost of owning machinery is so high that farmers often enter into complex sharing agreements. Some farmers share machinery 'across the fence' with neighbours, or amongst families within local areas. However the risk here is that the wheat matures and is ready for harvest at the same time. It is a race to strip the wheat before weather or other conditions begin to devalue the crop. Some machinery agreements (especially where farmers are adjacent or close by) mean that neighbours might do one paddock on farm A then one on farm B – each year alternating who goes first. Others, especially where distance separates farms, harvest all of one farm and then all of the other again alternating which farm gets 'first go'. The north-south lag provides opportunities for collaboration that solve these problems. A farmer in the north of the state will share the machinery with a farmer in the south, whose crop is ready six weeks later, thus guaranteeing that no-one is disadvantaged by having to wait to strip when their crop is ready. In some cases when farmers are looking to expand their holdings, they deliberately buy land at the opposite end of the wheat belt so that they can manage their time and labour by straddling the seasonal variation.

Pre-Sowing

Variety selection is an important part of strategic planning for growing wheat and it is approached differently by different farmers. Most farmers have an arsenal of three or four varieties prepared and ready in order to respond to the season as it emerges. The 'Winter Crop Variety Sowing Guide' is produced by the DPI. Attributes considered include grain quality to attract premium payments, good disease resistance, maturity suited to sowing time, strong seedling vigour, resistance to lodging and shattering, tolerance to herbicides, soil acidity, pre-harvest sprouting and frost, and good threshing ability (DPI 2010: 6).

Wheat is bought and sold principally as a conduit of protein, as measured by 'quality' standards. These are measured at various points in the post-harvest marketing chain, for example, the highest grade Australian Prime Hard requires protein levels over 13 per cent, while Australian Hard is over 12.5 per cent protein. (These standards are discussed more fully in Chapter 6.) Sometimes farmers chase particular quality bands – they grow wheat to try and get into a particular grade in order to get a particular premium price for their grain. Others focus on yield rather than particular quality parameters. The dance between these two, which interact in turn with moisture and nutrient requirements, is explained by Jack:

> We're trying to grow all [Australian] Prime Hard varieties. Because if we can achieve a Prime Hard grade, or quality, that increases our income. If we don't achieve the Prime Hard, we only come out with Australian Hard well, yes, our income is reduced. But not always, because we could have a yield advantage. Sometimes we don't get it right. We try to get our nutrient levels right ... If we get an exceptionally good season, even though we have a Prime Hard variety planted we may only achieve an Australian Hard quality ... we may not have put quite enough nitrogen to achieve that Prime Hard. (Jack, household D)

Most of the farmers we interviewed reported that they keep seed of at least two and often more varieties of wheat in order to be able to best respond to the season. Commonly they have one variety for an early break, one for a late break and one that can be sown dry. In addition they may have a variety with good rust resistance if there is a bad rust season predicted (usually indicated by what has happened in Western Australia during the previous season).

Sowing

Within the cropping calendar there are a few tipping points that drive crucial decisions (Appendix1). One is the arrival of 'the autumn break', the rain that determines the sowing. Negotiating the arrival of the break is a key point in farmers' descriptions of their year. It has been traditional to 'sow into moisture', meaning that they will wait for rain to fall and then begin sowing when the ground is dry enough to get the machinery on. Alternatively the break may never come and if you wait too long before you sow, the crop will not spend enough time in the ground to reach its full yield potential. In recent years where the autumn break has been very late or non-existent, some farmers have elected to sow their crops dry so that the seed will lay dormant in the ground until the rain comes. The risk is that if there is only a light shower, the seed may germinate but will wither and die without follow up rain. Ian describes these dilemmas:

> ... there's an ideal window for sowing your crop and if you miss that window or at the end of the window, basically you can't reach your full potential of yield. So it's better to probably gamble and put it in the ground when you've got a lot to do like we have and have it ready for the rain.

> Last year we sowed crop and we had no rain on it for two months. It sat there in the ground after we sowed it for two months and then it rained and it all just came up beautifully. It was quite happy. It's probably just part of the gamble of cropping where you've just got to start and put it in and hope to hell it rains. So you're just hoping it rains all the time. (Ian, household W)

Moisture profiles are generally more consistent in the northern part of the wheat belt where there is summer rainfall and soils hold moisture better. There are no set rotations used in this area, and farmers described themselves as 'response croppers', or 'farming the moisture'. One farmer described to us that his soil gave him more flexibility, meaning that the amount and timing of rainfall was not as critical as elsewhere. This was a key point of difference between the north and the south in their experiences of the most recent drought. While southern farmers experienced two successive crop failures, in the north yields were only slightly reduced and then sold into a premium market offering high prices.

The rate of sowing is an important variable in addition to timing. In the western margins of the wheat belt lighter sowing rates mean that each individual plant has less competition for scarce water resources and this results in better yields. Eastern areas with higher rainfall use much heavier sowing rates.

Feeding the Wheat – Fertiliser and Other Inputs

Australian farmers have long been criticised for their excessive use of fertilisers, pesticides and herbicides, which are blamed for eutrophication of waterways and other environmental damage. There is no simple way to produce a crop and nothing but the crop, and even organic methods have their problematic aspects, as we discuss. If it was ever the case that these decisions were made lightly, escalating prices of these inputs in recent years has certainly changed things. The complexity of these decisions, as well as the way that different generations deal with issues, is illustrated by our conversation with Gordon, a second generation farmer from southern NSW and his son Mark (household U). Mark left the farm to study agronomy at university, later working as an agronomist for a local reseller. He had returned to the farm to run the crop side of the business, with Gordon managing the sheep side. Gordon reminisced about conflicts with his own father Patrick over the use of lime, at a time when they had each run their own farms but worked together on large tasks such as harvest.

> I started using lime, I don't know what year it was, but dad saw everything as an expense didn't matter what it was, using chemical or something, they're expensive, he didn't see the cost benefit of it. And I'd been using lime here for quite a few years and I tried to talk him into using it no, no, wouldn't do it.

> Eventually I talked him into buying one load of lime and he ... got a contractor in to spread this lime, on one end of one paddock. Put a crop in it, and it was chalk and cheese. You could see where this contractor, [had] done a circle like that with the machine ... and in the crop you could see this circle. All of a sudden lime was the best thing he'd ever heard of. (Gordon)

In turn, Mark describes how he has negotiated the balance of inputs and yields, at the same time balancing academic training and practical experience over the last few years, with a bit of competitive behaviour thrown in.

> Mark: … well we were already getting five tonne crops there for a while and we thought well how can we go a bit better, how can we get that five and a half. And there was a fair bit of late urea put on, and that's how we got in trouble in [19]98, or [20]01 was the same with the frost.

> Gordon: We actually cost ourselves money because we over fertilised.

> Mark: Yeah and I guess there's been a bit of a swing in my thinking the last few years of not trying to aim for the big yields but keeping a reasonable average but at low cost. So I think we just got a bit carried away there for a few years.

> Interviewer: What prompted this swing?

> Mark: Finances. In a way. Probably there was a bit of ego in it I guess to start with. Chasing the higher [yield] … Yeah, and it was a district thing. I mean everyone was trying to do it, but I guess there's a lot of times there's no money in it. So it's just a matter of not so much looking at the outcomes all the time, but the costs in getting there.

Although northern farmers initially responded that the agronomics of durum wheat were essentially the same as bread wheat, most agreed that the nutritional management of durum was a bit more difficult and was the real key to reaching the top grades. Both nitrogen and urea were the key inputs to manage, and in general durum wheat requires more of these. Farmers explained that durum did not feed or 'scavenge' these nutrients from the soil as efficiently as bread wheat and so 'no stone was left unturned' in soil preparation.

By its very nature, production of a monoculture crop implies reductions in ecological diversity for, at the very least, the area of the farm. Organic cereal croppers attempt to manage the demands of plant growth using a different range of inputs, but broad acre farming presents quite different challenges to a permaculture vegetable garden. We interviewed a small organic wheat and cereal (oats and spelt) farmer in the western district of NSW (household H). Arthur described to us how he manages the farm on a rotation schedule with the livestock (a small number of cattle and sheep for meat) integrated into the grazing and fallow phases. The basic cycle has a crop followed by a rest period, another crop, and then seven years fallow. This is consistent with other accounts that organic agriculture operates on longer fallow phases than conventional cropping (Derrick and Dumaresq 1999). In this system organic farmers achieve much lower yields, but claim to be producing a higher quality product that commands a price premium five or six times higher than conventional wheat. This is a different assessment of 'quality' than is measured in the conventional Australian standards process.

The property is certified as organic with both the National Association for Sustainable Agriculture, Australia (NASSA) and Biodynamic Farmers of Australia (BFA), requiring that the only allowable inputs are those listed under the certification schedule. For broad acre cereal cropping this means for example that where phosphorus is deficient, it has to be sourced as rock phosphate, as opposed to superphosphate used in a conventional system. Arthur told us that he relied upon soil microbes to make the phosphorus in rock phosphate available to the growing crop, considering superphosphate to be an unnatural and artificial product which delivers nutrients to the plant in an unnatural way. (It should be noted that the input substitution used by organic cereal farmers has been argued to be a key element in the erosion of the principles underlying organic agriculture (Lockie et al. 2006).)

The common practice of tillage rather than use of herbicides in organic cropping is another major distinction from conventional wheat cropping, which routinely uses herbicides both before and after crops are sown. A wide range of broadleaf weeds compete with the growing wheat plant for water and nutrients, and can potentially contaminate harvested grain with their seed. In contrast, Arthur considered weeds as indicators of soil health, identifying different deficiencies according to whatever weeds may be present, and adding trace elements where he thought necessary. Where weeds are present in his crop they must be removed without the use of chemicals, usually by physical removal using disk ploughs which were described as an 'old fashioned' farming practice, similar to how people farmed before herbicides were available. The soil is cultivated using the disk plough during the summer, a process which also breaks up any clods in the soil, and then during sowing the soil is again cultivated. Arthur explained to us that when they first started farming they set the cultivation tines quite deep, but have learned through experience now to cultivate at a depth of only a centimetre or two.

The major drawback of this system is that the soil profile is disturbed, and potentially more prone to both wind and water erosion. The now conventional no-till or conservation-till systems have had an enormous impact on rates of soil retention since their introduction and development in Australia in the 1980s (Silburn et al. 2007). Although there have been important advances in organically based tillage technology, using weed sensors with minimal soil disturbance, these technologies are expensive and size efficiencies are required. Even with the premium prices available for organic wheat and spelt, this was not considered to be a worthwhile investment by Arthur.

Harvest

> And so October and November it's just approaching harvest, so you're just checking for diseases and doing yield estimates, and then just organising the harvest operation and making sure that the machines and that are all ready for harvest.

> That's a very busy time of year and you get adrenaline running and everybody who grows crops are all doing the same thing at the same time, so once harvest commences you're just flat out harvesting trying to get the crop off before any rain or storms … Once you get a certain amount of rain on it, it starts doing quality damage which costs you money so that's why you're always keen to get it off as soon as you can. (Joseph, household G)

Harvest time is frantic and busy. Due to the drought, we could only hear stories of the excitement rather than being able to observe it in the southern part of the study area. We were able to see some harvesting in the north (Figure 5.5). A few days here or there can make big changes to the quality of the wheat. Whereas during the rest of the season there is almost no such thing as 'too much rain', come harvest time farmers cast a worried eye over every passing summer storm cloud. During one field session in the north a large rainfall event and subsequent flood took place, reported by local residents in the media as the largest flood they had experienced in 30 years (Belt and Sheridan 2008). Some farmers in the district, especially in lower lying areas suffered substantial flood damage – debris and water over the near ripe crop (Figure 5.6) culminating in the area being officially declared a disaster zone (ABC 2008). The more insidious problem of rainfall during harvest, however, was that the crop over a much wider area was damaged and, in severe cases, shot and sprung in the paddock, severely downgrading its value.

Figure 5.5 Harvesting is a frantic and busy time of the year

Some farmers elect to use contractors for harvesting (also referred to as stripping), rather than incur the expense of machinery ownership or leasing. However this means a loss of control and independence. When the harvest is on everyone wants to do it at the same time and contractors, who have to make their entire annual income in a few short weeks, often over-commit themselves and then arrive late to strip the crop. Harvesting their own crops also offers farmers the advantage of closely observing their paddocks (Figure 5.7), as James explains:

> ... they get to see the fields and they know that section went well, this section didn't go well, what happened there. They can join the dots through the season. So the good ones will probably spend a lot of time in the header, if they can. It's not always logistically possible ... There's a lot of variation within fields ... and it's not always easy to do that, but if you've seen it the whole way through you're more likely to be able to diagnose problems and implement plans for the future. (James, private agronomist)

Figure 5.6 Flooding along the Namoi River and low lying wheat crop, northern NSW, December 2008

Figure 5.7 Harvest from inside the header

Storage and Innovation

Farmers often aim to increase their resilience to externalities by buying protective infrastructure, such as silos or grain storage technologies. By enhancing on-farm storage, they can choose when to sell, avoiding both the flooded market and busy-ness and stress of harvest time. As well as permanent storage facilities, there is increasing investment in silo bags or other temporary storage facilities such as under canvas or in sheds. Silo bags are a low cost polyethylene bag developed in Argentina which can store up to 300 tonnes in an airtight environment (Lawrence and Caddick 2006) (Figure 5.8). This was described as giving more flexibility, particularly in choosing when to sell, so that they were not 'ripped off' at harvest time when prices were usually poorest. On-farm storage also aids general farm management, particularly in the north where harvest coincides with sowing and attending to the coming summer crop.

Figure 5.8 Silo bags used to store grain on farm

Most of the farmers were very aware of the new and different technologies available, but the huge capital costs involved prohibited them from innovating as fast as they might like. Dave, a durum farmer, had invested in extensive on farm storage facilities which also gave him the ability to blend grain on farm (Figure 5.9). Grain could be segregated into four different stockpiles of varying quality attributes, and then mixed in a mixing silo to meet the particular specification that was to be sent to a receiver. Better quality grain could be mixed with poorer qualities to meet a higher overall average. Dave explained that his strategy was to 'evolve' into new technologies as each could be afforded, but that did not necessarily mean waiting for old machinery to fall apart. He also said that in some cases he had only realised some of the different advantages available from new technology after he had bought it. Other farmers had pooled resources amongst neighbours and family members to take advantage of new technologies.

Figure 5.9 On farm storage and mixing system, northern NSW

Another technology being utilised in northern NSW was aerated storage, where grain that was too wet to be delivered to the market could be dried down to the required moisture levels. This technology came into its own during the wet harvest that we observed, allowing farmers to continue harvesting grain that might previously have been left in the field to dry out and become further damaged. One farmer was using this technology to harvest grain earlier than others in the district, and had avoided most of the rain during his harvest. Other recent technologies in which farmers had invested were caterpillar tractors, which minimised soil compaction over regular wheeled tractors and also 'auto steer' linked to satellite navigation, which reduced chemical usage by minimising over-application during tractor turns in the paddock.

In some of the interviews, farmers spoke about the idea of a 'race' to adopt technology and the need to be seen as innovative. But this is a careful balancing act. There was awareness that new technologies have little bumps and challenges, so farmers do not necessarily want to adopt too early and suffer during the process of ironing out the kinks. One farm aimed to be 'early adopters' with technology.

> We're being early adopters. We're trying not to be the bleeding stage but that has happened a bit with the disc seeder last year … they hadn't been developed for places that had the heavy stubbles we had, but we've made a few modifications.

We've worked very closely with this guy ... an agricultural engineer, and there's
not very many of them in Australia. We're pretty pleased. (Janice, household X)

Many farmers talked about a paradigm shift in the way that knowledge is now
gained and transferred. Whereas the last generation were quite secretive about
their farming habits, people felt there has been a transition towards new kinds
of agricultural extension and a greater degree of knowledge sharing across the
farming community.

After harvest there is a bit of time to reflect on the mistakes and successes of
the past season, and to decide on the actions that will be taken for the next one.
With harvest fresh in their mind, farmers have ideas about areas where fertiliser
is required or where rotations need to be altered due to rainfall or disease. They
begin to work out how much grain they will retain for seed (usually the best and
cleanest part of the crop), and which varieties they will use. They also consider
what kinds of inputs will be ordered and how they will manage the stubble of the
previous crop. They might be marketing their crop from last harvest or taking out
insurance on next year's crop or managing their hedging instruments. They might
even take a holiday.

Drought and Time

All interviewees spoke extensively about the drought, and all understood
drought as a normal and expected part of farming in Australia. There is no
single experience of this process, since, as Vince (household T) said, 'When the
drought starts ... it just happens really slowly and the dams go down slowly and
everything, and you sort of adjust with it'. We have examined those interviews
conducted between December 2006 and May 2007 for the extent to which the
current and recent droughts were characterised as 'normal' or indicative of bigger
climate changes. Several farmers were inclined to think of the present conditions
as being part of normal variability (Appendix 2). More thought the 2000s drought
was significantly different, because of its length (between three and seven years
depending on location) and intensity. The latter is expressed mostly in long term
depletion of soil moisture levels. This group includes the oldest study participants
who have personal memories of the 1940s drought (households H, K, P, Q, S
and V). Stories of earlier droughts also become part of the climate memory of
younger generations, who have listened to grandfathers and other older men over
the years. On two of these older properties (H and Q) we were shown family logs
that provide long term documentation of rainfall records. Farmer perceptions that
the drought we observed was significantly different are supported by the changed
practices that they describe (Appendix 2). These included having to cut and sell
wheat for hay rather than grain, and having to buy in seed for next year's planting.

Although more farmers than not considered the current circumstances to
indicate a significantly different drought regime or a changed normality, this was

not generally attributed to climate change. In the 24 interviews undertaken during this period, climate change or global warming was mentioned in seven. In two of the seven the drought was attributed to long term climatic cycles rather than anthropogenic climate change, and in two there was attribution to climate change, albeit with some uncertainty. For example

> When you listen to all the media and global warming and the changing environmental issues right around the world, maybe there is the possibility that things have changed so radically that perhaps the big rain events historically may be a lot further away for us. So the confidence for me has gone to a large degree. (Vince, May 2007)

The three other mentions of climate change related to the likelihood of a changed economic environment, such as the possibility of getting carbon credit for tree planting, rather than to weather, drought or climate per se.

The growers in the north had a rather different perspective on drought. Those interviewed all thought the Liverpool Plains were one of the most fortunate farming areas in the country. When asked about drought and poor seasons in their business, each grower replied that they did not really think they had had a drought in their region. None of the growers could remember a year when they had not harvested at least one crop, and most years both summer and winter crops had been achieved. Two factors explained their good fortune, the summer rainfall on the plains, which supplements and at times compensates for lack of winter rain, but more significantly the ability of soils on the Liverpool Plains to hold moisture.

> I'm aware of how temperate region droughts affect [them], and particularly when you're in once a year cropping, monoculture type things, you know, the impact it has is pretty dramatic. I'd have to say that that's not the effect here. We might have poor yields but, I mean, I actually quite enjoy droughts. I mean, the pressure's not on to do things. We make very good profits – even though we have a lower output in terms of tonnage, our costs are so much lower. We actually are still producing something, its value is quite high. Our taxable income goes up often and as I said, I quite enjoy my droughts. I know they're debilitating in many rural parts of Australia. (Zack)

Risk: Strategic and Reactive

Studies on risk and danger distinguish between external (= analytical, scientific, expert) and internal (= affect, individual) definitions (Dessai et al. 2004, Lowe and Lorenzoni 2007). These researchers emphasise the human elements of risk, and draw on the concept of the social amplification of risk, when different perspectives and types of risk interact with one another. The external and internal perspectives are paralleled in the psychological literature on risk by analytic and experiential modes of thinking (Slovic et al. 2004).

Participants discussed drought as a risk that farmers must manage, and also as a phenomenon that goes far beyond climate. Risk emerged in our data as a recurring theme. Farmers and their households are au fait with the language and experience of risk, and talk about a number of different risk domains. However there were differences in the extent to which study participants felt themselves victims of risk, or embraced it as providing opportunities. We summarised these as *reactive* and *strategic* respectively (Head et al. 2011). We recognise that individuals or households may not fall neatly into these categorisations; few are either totally strategic or totally reactive. The categorisations were developed empirically from the interview data. There are intersections with the way adaptation and risk have been understood and categorised in other agricultural studies (for example Bryant et al. 2000, Smit and Skinner 2002, Bradshaw et al. 2004, Reid et al. 2007).

Strategic approaches emphasise the active role of the farmer in negotiating risk. Sometimes this is to the point of seeing new opportunities in difficult circumstances. Reactive approaches see risk as being out of their control, and tend to position the farmer as a victim. For example, several farmers used gambling as a metaphor for their cropping experiences. While this is sometimes an example of laconic humour used by even the most successful farmers, it emphasises that some farmers feel vulnerable and perceive as unmanageable the array of risks to which they are exposed and which they must negotiate.

Our interview data shows that different types of risk interact, and tend to coalesce into patterns of vulnerability (reactive approaches) and resilience (strategic approaches). Household W in the south of the study area, provides an example of the latter. This farm has part of its cropping area under irrigation due to possession of a license to tap into an aquifer. Because of its location, household W has many options for marketing wheat, including to the stock feed mill not far away. Household members are well travelled, for example they had been to Argentina to look at innovations such as silo bags for on-farm grain storage. Succession planning has been very deliberate in this family, with three adult children all tertiary educated in agriculture or business related areas. They talk about risk in the following ways:

> and the weather is just a risk that you've got to manage because that's just what happens if you're farming. You never ever think what's going to happen long term with the weather, so you just factor that in and just hope for the best ...

> So even though we've had to borrow a lot of money to develop it [irrigation farming] it's sort of a risk management tool from a business point of view because we're constantly aware that you can have droughts.

> Farming … you just try to constantly manage risk, that's really what we're doing
> nearly all the time. You have to think of the worst case scenarios and how we
> can best try and deflect those. We won't be able to stop them but you try and
> deflect them. Sometimes it works and sometimes it doesn't work as well. (Ian,
> household W)

In contrast Household N, in the north west of the study area, described various
economic struggles, such as not being able to borrow money to replace old
equipment. The husband drives trucks off farm to help pay the bills. At the far
western margin of wheat viability, household members openly discussed having
cleared some areas of native bush, ahead of legislative prohibitions, in order to
expand the cropping area. The three adult children would like to continue working
on the farm, which at the moment only supports one of them. Another son has
completed an apprenticeship and will take whatever labouring jobs are available
to make ends meet, while assisting his parents on weekends. Susie, the mother,
articulates a very different sense of risk to Ian:

> I always laugh when we go to church, they condemn gambling, I'm thinking
> farmers are the biggest gamblers out … Like you borrow money, you borrow
> money, thousands of dollars and stick it in the ground and keep your fingers
> crossed that in eight months down the track you're going to get something out of
> it… we may as well plant our money in the ground because at least at the end of
> the year we can dig it up again. (Susie, household N)

Although it is not a neat and tidy process, approaches to drought risk interact
with approaches to marketing. Another important variable is geographic location.
Farmers further west not only have lower yields due to less reliable rainfall, but
also fewer marketing choices because of higher transport costs. Where the farm
has been struggling for many years there has been less opportunity for the next
generation to gain tertiary qualifications that would strengthen the business. In
the east and south, several farms have adult children with tertiary qualifications
in agriculture, agronomy, economics, and finance, who have entered the family
business in different ways. This is not to say that the demise of the former is
inevitable, but that its sustainability and security is much less assured.

Risk: Affective, Embodied and Everyday

The different dimensions of risk management extend into many aspects of daily
life. The affective or emotional aspects of risk take a variety of expressions. As
a result of the deregulated economic environment and its global teleconnections,
daily life on all these farms now involves engagement with much more than the
weather and the tractor. Both strategic and reactive farmers described complex
juggling of information. In this process farmers need to be just as tied into global

communication networks as any urban office worker, but they also have to maintain a daily paddock life with one eye on the weather. Farmers varied between feeling swamped by this and accepting it as part of the skills required to do business in the present day. Older farmers such as Arthur grimaced at the blizzards of paper providing demands and information: 'I can't get over the material I'm supposed to read and when it [the paper pile] gets about that high, I haven't had time to read it, I put it in the waste paper basket and start building another one'. Jim (household M) contrasted the current situation with 'the old days' when there were fewer decisions to be made: '... the old days, you just cart your wheat to the silo and get whatever money. You don't do that anymore. You're just on a mobile phone all the time ringing up, see who's got the best money for just about every load.'

The young and business-savvy Chris (household B) provides a rare example of relative comfort with the constant switching between farming and office modes of work: 'Hop on the internet every morning, have a look, see what the market's doing ... you get the live, 15 minute delay on the internet. See what the market's doing and ring up the broker and he'll tell you. It's all pretty easy ...'

A further instructive example is the emergence of the new technologies of on-farm grain storage as an important generic risk management strategy. As we saw above, on-farm storage enables the farmer to control the timing of selling, rather than having to sell everything at harvest when the price is lowest. In a broader context that now includes a deregulated market, many farmers have in effect become their own grain traders. This is burdensome because it increases the amount of time spent on the phone and the internet, relative to time spent in the paddock. In some larger farming households, one person can specialise in the business and marketing side, but this is rare.

Approaches to risk are not immutable, and not solely an outcome of individual psychology, but they do have expression in individual bodies. This is seen particularly in relation to the multiple risks entailed in forward selling part of the crop. During the early part of 2007 a number of farmers had been encouraged by their banks to forward sell, in anticipation of a good season. Peter and Heather (household Y) spoke honestly about their experience of being burnt by forward selling. Heather's view had been:

> We had a little discussion earlier in the year which is quite amusing and we talked about forward selling and I said, "oh look, I don't like it." It seems to me that whenever you do this forward selling thing half the time you're better off and half the time you lose money on it, seems to me why put yourself through the stress you should just sell it when you've got it.

As we mentioned in the introduction to the chapter, Heather also discussed how the stress had caused Peter's physical symptoms such as headaches and sleeplessness. He was not the only one in the district having these experiences.

> I think also, like besides even the money lost, it seemed to cause an enormous amount of stress through all the second half of the season really, when it stopped raining, because they were all really worried about whether they were going to be able to fill their contracts. Our neighbour who's a fabulous farmer and his crops are looking beautiful and I was saying 'Oh you know, your crops are really holding on', and he said 'Oh look, if I hadn't forward sold so much I wouldn't be worried at all' ... He was not sleeping at night was he, and really under a lot of stress because of the forward selling. (Heather, household Y)

Peter's physical symptoms are just one illustration of how the drought/climate change/financial assemblage can have pervasive outcomes for wellbeing within households. For farmers like these, adaptation to climate change may in the end be measured by evaluation of changes in quality of life, rather than a raw economic or environmental viability decision.

Conclusion

Marketing was another particularly important issue to farmers. As we noted in Chapter 4 the Australian market for wheat had been deregulated in the previous 12 months and wheat farmers were able to experience for the first harvest how this might impact their business. Some farmers thought that the Australian Wheat Board (AWB) had successfully sheltered them from the substantial economic risks which occurred at harvest time, while others thought the main problems that had arisen happened because the full price advantages had not been transferred back to growers.

The issues of marketing and deregulation highlighted the wider issues of risk and uncertainty in Australian wheat farming. Some farmers repeatedly spoke about uncertainty and the changes they experienced from year to year. Conditions experienced one year could be completely reversed the next.

> Yeah, you'd run for the hills because you've not only got the risk of your costs moving all over the place, you've got the risks of your income moving all over the place and then on top of that you've got the imponderable and variable weather conditions that can destroy a year's work in, in 20 minutes. (Dave)

We come back to these specific themes of risk and uncertainty in the context of climate change in Chapter 9.

Chapter 6
Mobility, Friction and Fungibility

Vacant and dusty flour mills dominate the landscape of many country towns in central and western New South Wales, an emblem of so-called 'dying town' syndrome (cf. Worster 1979). Now being assessed for their heritage significance, they and their associated silos are the material remains of a more fine-grained rural life than exists today, the kind of life grieved for by the farming communities interviewed by McCann (2005). Sometimes derelict, sometimes restored to a new life such as in the form of an organic liquorice factory at Junee, the present fossil status of the mills and silos seems to indicate stasis and immobility. They are the strandline behind a modernist wave of mobility and dynamism that has stripped jobs out of agriculture and sucked youth to regional centres and capital cities.

A closer look at wheat through the lens of mobility shows the mills to have been points of both movement and friction; indeed they illustrate that movement and friction are two sides of the same coin. From the moment of harvest wheat is on the move, but it must also be stored at various points in time and space. Friction in this flow is expensive – it costs time and money to load and unload trucks, trains and silos.

In this chapter we draw on recent social science scholarship on mobility. This is sometimes referred to as constituting a 'new mobilities paradigm' (see for example Sheller and Urry 2006, Cresswell 2010), that has sought to engage with the enhanced and complex mobility of the modern world. The most useful of these writings argue that 'the *complex* character of such systems stems from the multiple fixities or moorings often on a substantial physical scale that enable the fluidities of liquid modernity' (Sheller and Urry 2006: 210). The hyper mobility of air travel, for example, is facilitated by a very immobile place, 'the airport-city', from which tens of thousands of workers orchestrate the movements of others (Sheller and Urry 2006: 219). Bissell (2010) draws on Latourian ideas to consider the systems of maintenance, repair and administration, that is labour, that go into the production of a smooth train ride. Thus we can understand that 'mobility is one of the major resources of twenty-first century life', the differential distribution of which is an important political question (Cresswell 2010: 22). Similarly, an enormous investment of labour and materials enables wheat to become mobile and to flow in diverse ways.

Anthropologist Anna Tsing (2005) uses the concept of *friction* to describe the 'sticky engagements' of global connections. On the one hand Tsing contests simplistic depictions of modernist mobility in which 'the flow of goods, ideas, money, and people [is] pervasive and unimpeded' (2005: 5). In fact, she argues,

insufficient funds, late buses, security searches, and informal lines of segregation
hold up our travel; railroad tracks and regular airline schedules expedite it but
guide its routes. Some of the time, we don't want to go at all, and we leave town
only when they've bombed our homes ... (Tsing 2005: 5–6)

On the other hand, 'friction is not just about slowing things down. Friction is
required to keep global power in motion ... Friction inflects historical trajectories,
enabling, excluding, and particularizing' (Tsing 2005: 6). Friction can be creative,
as it is in the northern NSW town of Gunnedah, where the Namoi Flour Mill is
located less than two blocks from the town centre, adjacent to the intersection
between the rail crossing and the main road. One of the large town churches is also
located directly adjacent to the mill grounds. The activities and sounds of the mill
are part of the activities and movements of the town's people as they manoeuvre
around freight trucks and wait for rail haulage to pass at the crossing. Contrasting
with the silent mills of rural landscapes in decline, for this town friction constitutes
proof of economic life.

Storage and stillness can usefully be thought of as mirror concepts to mobility.
As Bissell argues in his thinking around the passenger sitting in a train, tracing
vibrations 'opens up a way of thinking about the uncertain and provisional
connections between bodies, their travelling environments and the experience
of movement where movement is not opposed to stillness' (2010: 2). We need
to think about when the wheat stops moving, and sits still. These moments of
stillness can also be times of vulnerability to attack of pests, humidity and fire in
silo, train carriage or road tanker.

Most of the new mobility thinking has focused on people rather than things
other than human (Cresswell 2006) but, as Pawson (2008) has argued, this literature
can be brought into conversation with a longstanding tradition of geographic and
historical writings on the webs and flows of plant exchange across the globe (Sauer
1952, Crosby 1972, 1986). Pawson aims to cut across the unidirectional flows
implied by Crosby's title *Ecological Imperialism*, showing that plant movements
in modernity have not all diffused from a colonial centre. Rather, a closer look at
the evidence shows webs and networks of exchange in many directions. Pawson
extends agency beyond the European actors such as botanists Banks and Hooker,
who have received most of the research attention, to indigenous and vernacular
people, for example, throughout Australasia and the Pacific. He also implicitly
acknowledges the agency of the plants themselves in his discussion of prickly pear
becoming invasive in Madagascar, and Australian orchids colonising New Zealand
across the ocean. In a further knotting of the web, Australian figs introduced by
people 'have become weeds around Auckland since the airborne arrival of the
Australian wasps that pollinate them' (Pawson 2008: 1472).

This chapter examines a number of moments in the mobility of wheat, mostly
after it leaves the farm. For the most part, only part of the plant – the grain of
wheat rich in starch and protein – is moving. The process of movement is enabled
by a complex network of people, machines and information. Without denying

the breathtaking scale of contemporary mobilities, the narrative moments around which this chapter is structured emphasise that these theoretical understandings of mobility can be applied productively to previous times and places. We also use the lens of fungibility to examine the way wheat's identity changes over time and space, becoming less or more differentiated in different contexts.

The generative possibilities of friction and vibration in the work of Tsing and Bissell respectively help us reflect not only on the mobility of wheat, but also on ourselves as mobile researchers, who moved 'along with the people, images or objects that are moving and are being studied', attempting to capture 'through observation, questionnaires, interviews, mapping and traces, the complex mobilities of the people, images and objects under study' (Larsen et al. 2006: 6). We moved across large distances, in and out of cars, trucks, headers, kitchens, offices, paddocks and planes. We were sometimes alone, sometimes together. We could organise certain things in advance, but had to be very much open to the moment, as Jenny's story below illustrates. The process of research is not smooth, indeed it can be the frictions and vibrations that generate the most interest.

Fungibility and Mobile Identity

A fundamental concept in the movement of wheat is *fungibility*, the property by which individual units of something are mutually interchangeable. Wheat, like money or oil, can be and is replaced in the trading system with another 'identical' item. This term comes from the Latin fungi, 'perform, enjoy', with the same sense as to 'serve in place of' (Stevenson 2010). Wheat can stand for other wheat; it very rapidly loses its identity as an individual plant or plant part. We asked a grain trader in suburban Melbourne, 'where is the wheat?' He replied, 'It doesn't really matter, the wheat is all out there', and proceeded to describe a system of receival and bulk handling at once virtual and pinned to the material realities of moving grain. The receival places in which his company rents space 'are all the major silos that you'll come across when you are driving around the bush'. In this process many buyers and sellers are circulating the same wheat: 'everybody's product is co-mingled. So it is not identity preserved, if that makes sense. That means that you and I and five other people all buy wheat into the same stack and we take out whatever the average of the stack is' (Malcolm, grain trader, 2007). In contrast Terence, a keen entrant in wheat shows such as the Sydney Royal Easter Show, prepares his entry by picking through the best of his bulk harvest grain by grain (Figure 6.1). He uses tweezers and paintbrushes to sort the wheat into boxes and jars according to characteristics such as size, fullness and colour. Each of the boxes then contains as consistent a sample as possible according to qualities that are scarcely detectable to the lay observer.

Figure 6.1 Picking out single grains for judging at the Sydney Royal Easter Show

An important historical moment in the co-mingling of grain was the shift from bagged wheat to bulk wheat transport, as told by Cronon (1991) for Chicago, and echoed in western New South Wales in the early twentieth century. Development of Australian standards underpinned the development of the bulk storage and handling system, since 'agreed receival standards provide a basis upon which buyers are able to verify that the wheat they receive matches the description of the wheat they have purchased'(Productivity Commission 2010: 259).

The intensification of mobility over the last century or two may suggest that any identity has been completely lost in the undifferentiated flows of wheat circling the globe. On the contrary; one of the themes of this chapter is how new identities of wheat are being created and differentiated through different qualities and standards, so that Malcolm has many more different stacks to choose from.

As we saw in Chapter 5, farmers have planted particular varieties of wheat, aiming to optimise the agronomic conditions of their region and farm, and target particular markets. Varieties are divided into different classes, for example APH (Australian Prime Hard), AH (Australian Hard) and ADR (Australian Durum). These classes determine the maximum grade into which a variety may be received. Many factors influence the quality of the harvested grain up to the point of receival, which is where its grade is determined. For example, there are two grades of Australian Prime Hard Varieties (APH1 and APH2), and three of Durum wheat (DR1, DR2, DR3) (Table 6.1). The standards for each grade are an interesting mix of chemical (protein content), botanical (seeds of weeds or other crops) and physical (moisture content, presence of gravel) parameters which are each quite strictly defined and can be measured in a sample taken from the truck or train. Wheat that does not meet the standard can be sold into the lower grades via the 'Bin Grade Cascade' (GTA 2011a: 46) which, despite its beautifully flowing name, is more of a slippery slope, since all Cascades end in the dreaded FED1 – feed wheat. In other situations the wider context can completely alter the basis on which the wheat is sold into the pool; it effectively has to have its identity changed to get a sale. The process and its tense intersections with mobility are illustrated by Jenny's afternoon hauling grain with Graham at Sunny Hill.

Table 6.1 Common grades of Australian wheat

AGP1	Various Varieties except FEED (General Purpose Grade)
ANW1	Australian Standard White Noodle Varieties
ANW2	Australian Standard White Noodle Varieties
APH1	Australian Prime Hard Varieties
APH2	Australian Prime Hard Varieties
APW1	Australian Premium White Varieties
APW2	Australian Premium White Varieties
APWN	Australian Premium White Noodle Varieties
ASW1	Australian Standard White Varieties
AUH2	Australian Hard Varieties (Utility Grade)
AUW1	Various Varieties except FEED (Utility Grade)
DR1	Australian Durum Varieties
DR2	Australian Durum Varieties
DR3	Australian Durum Varieties
FED1	Various Varieties (Feed Grade)
H1	Australian Hard Varieties
H2	Australian Hard Varieties
HPS1	Australian Hard Varieties (High Screenings, High Protein Grade)
PNC	Cadoux variety
PNE	Eradu variety
PWT	Australian Korean Noodle Blend Varieties
SFE1	Australian Soft Varieties
SFT1	Australian Soft Varieties
SFW1	Various varieties (Stockfeed Wheat Grade)

Source: GTA 2011a: 8–9.

She was invited after an interview with Graham's boss Dave to look at his grain storage and mixing facility for durum wheat. It was the first thing she had noticed when she drove onto the property; a dozen or so massive white and aluminium silos connected by a spidery network of chutes (Figure 5.9). A weighbridge, unloading bay and basic testing facility lay on the far side, and a reloading bay sat underneath the central suspended mixing silo. As Jenny tells the story:

> This was a big installation, by far the biggest thing we'd seen so far, and I imagined it was pretty expensive. But it was not the only one in the district, and Dave had said during the interview that it had paid for itself pretty quickly. Originally he'd bought it for his sorghum crop – ironically at that time you'd get the best price for a crop 12 months before it was planted. These silos and the mixer gave Dave more control over the price he obtained not only because it gave him flexibility about when to sell, but also because he could separate and remix his own grain to achieve an overall better price, most of the time.
>
> Graham had his trailer parked under the mixing silo and stood, finishing a cigarette, waiting for a load to fill. Like all Dave's employees, Graham wore regulation safety fluoros and dusty King Gees. Evidently he'd taken a few loads out already that morning. As we stood admiring the silo, Dave introduced me to Graham who immediately offered to take me over to Sunny Hill in his truck. Sunny Hill is the main GrainCorp facility in the district and the destination for most of Dave's grain that week, about 16km away on the dirt road. Graham reckoned about an hour and a half return trip. I did a few silent calculations – not absolutely sure it was wise to accept a ride in a semi-trailer from a complete stranger in a semi-remote area. But I concentrated on the fact that I wouldn't get such an opportunity anytime soon, and was relieved as Graham fired off a rapid succession of questions, facts and notes of interest as we started out. It only took us five minutes to work out we'd attended the same country high school, although not at the same time, which seemed to reassure Graham that I wasn't too weird '… out here on my own, keenly interested in his truck and the wheat …'.
>
> After half an hour or so on the road and with the GrainCorp Silo in sight, Graham pulled up at the rail crossing to stop for a Manildra train. We waited for ten minutes as the thirty or so carriages pulled past painfully slowly, and Graham said that this was ominous. Apparently we could be here for hours, and there were now three or four other trucks queued up behind us, so there was no way of doubling back and getting around the train. Graham made a quick call on his iPhone to someone familiar and seemed relieved. This train was just going through, not receiving a load, so we shouldn't be too much longer. We crossed the tracks and stopped at the weighbridge (Figure 6.2). Graham explained that we'd be here for a few minutes as they tested the quality parameters of the grain, so I may as well get out and come and have a look at the small laboratory (Figure 6.3), which sits directly above the bridge. Graham's wife Sheila operated the weighbridge. She was, it turned out, who Graham had just phoned. After another ten or so minutes Sheila informed Graham that the load was not up to spec.

Dave had instructed Graham he was aiming for DR1 or DR2, all specs fine it seemed except the 'falling numbers' weren't right. The load checked out at DR3. This was not a great situation, a lower grade and hence a lower price than Dave was anticipating. A quick phone call to Dave to check what to do. 'We could bring it back and remix it.' With Dave's massive storage capacity it was possible to bring a load of grain back and remix with a portion of higher quality grain to achieve a better overall grade. The double handling and loss of time would still be worth it if Dave could get a better grade. But right now harvest was in full swing. The district had received record rainfalls and flooding the previous week, and Dave had to get the grain out of the paddock immediately so it didn't spoil any further. The poor falling numbers in this load were a direct result of that recent rainfall, but the risk was that if Dave delayed getting the rest of the crop in quickly, that recent moisture would result in the rest of the crop being 'shot and sprung' in the head, making it only fit for stock feeding, a lower grade again. Dave didn't have the immediate storage capacity to remix at this point. A quick decision – there were trucks banked up and waiting – 'Send it in, we don't have the space.' Graham did six return trips from Dave's place to the silo that day.

Figure 6.2 Waiting in the truck at the weighbridge, Sunny Hill

Figure 6.3 The weighbridge testing laboratory

What was going on in the 10 minutes that turned the wheat from the hoped-for DR1 or DR2 into DR3, and risked FED1? A minimum of three vertical probes drew a sample comprising at least 3 litres from each bin load. The sample was mixed, then separated to be weighed, sieved for defects and contaminants. Moisture and protein assessments were made 'by an appropriate field instrument calibrated against approved reference methods' (GTA 2011b: 1). The problematic Falling Numbers arose from a quality test to measure the amount of weather damage. The test goes to the plantiness of wheat. It

> is based on the unique ability of alpha amylase (an enzyme released during seed germination) to liquefy a starch gel. Strength of the enzyme is measured by Falling Number defined as the time in seconds required to stir plus the time it takes to allow the stirrer to fall a measured distance through a hot aqueous flour or meal gel undergoing liquefaction. (GTA 2011a: 6)

Graham's truckload would not have reached the Falling Number of 300 seconds to be graded as DR1 or 2, even though its protein specifications may have achieved the required 13 per cent (DR1) or 11.5 per cent (DR2). It 'cascaded' to DR3, for which it must have met the Falling Number of 200 seconds, thus avoiding a further cascade to FED1, for which there are no protein or Falling Number specifications.

(Note here that ending up with feed wheat prices for Durum wheat is especially costly due to the additional inputs required in the growing process. For some farmers located close to stock feed mills, the lower transport costs to the point of receival may make feed wheat the grade of choice. Indeed at the height of the drought, feed wheat prices were quite lucrative.)

At hundreds of receival, transfer or intersection points across the wheat belt and its transport network, the identity of wheat is being mediated and translated in similar small laboratories against the same set of standards. These are points of friction in that they temporarily stop the flow, but they also facilitate and orchestrate it, determining the precisely marketable streams of wheat with which each bin or truck load is fungible. At the Port Kembla export terminal two laboratories operate, with an automatic sampling system, to test grain before it is unloaded from trains to the silos, and again prior to loading onto ships. The second lab tests grain live as it goes into the ship, with Australian Quarantine and Inspection Service (AQIS) staff checking for insects at this point. If insects are detected during loading, quarantine inspectors instruct the terminal to cease loading from the infected bin and source grain from a different bin. From country silo to ship, Manager Matthew commented to us that the grain has been tested four or five times, so there is not a lot of room for surprises once it is on board.

Figure 6.4 Pardey's Mill, Temora, 2007

Pardey's Mill, Temora

Things were different when wheat had to be bagged, as illustrated by the history of Pardey's Mill, a prominent landmark in Temora, a town that celebrates its wheaten heritage with a local museum supported by an active group of volunteers (Figure 6.4). (This section draws on historical research by Gates 2009, wherein detailed sources can be found). With the establishment of wheat in the Temora district in the 1870s, and the town itself in 1880, it was only a matter of time before Temora needed flour mills. Pardey's was one of several in the town, in its present incarnation built in 1908.

This 'new' mill comprised three floors and a basement, along with a large grain shed that had a capacity of 40,000 bags of grain. The 'well-known metropolitan builder' Mr Dunkley mobilised a considerable suite of resources in the construction of the mill; 350,000 locally made bricks, hardwoods from the northern rivers of NSW, milling machinery from Thos. Robinson & Sons, Rochdale, England and a more than 100 horsepower 1907 Robey steam engine produced by Chapman & Co., Sydney, with steam supplied by a large wood-fuelled boiler. The water needed to run the steam plant became a problem as Temora had no permanent water supply, so at times water was carted in train loads from Gundagai and Griffith.

The top floor was a store for the total products of the grain after handling by the machines in the lower floors. There were three elevators, one being for flour. As the bag reached the correct weight, the machinery was automatically thrown out of gear, allowing the bag to be sewn and branded. The second and third elevators were for bran and pollard respectively. In a 1908 interview with the 'Temora Star', Arthur Pardey explained the workings of some of the mill machinery. Attention was drawn to the scourer that cleaned the wheat, in which it was 'rattled, sifted and blown' until every particle of dust and dirt had been removed. Next was the oat cylinder where all the stray oats or barley were separated from the wheat. There was also a dust collector that looked like a huge revolving wheel. Its aim was to collect all the dust blown out of the wheat by various machines to improve the atmosphere in the mill.

On the first floor were the four machines (rollers). The first roller started the first break of the gradual reduction system 'where nothing is done suddenly'. After the wheat was coarsely broken it was elevated to the top of the building where, by means of silk screens, the flour was taken out and the residue sent back to be crushed again by a finer set of rollers – the second break. Mr Pardey put his hand in a small aperture in one of the machines to pull out a handful of what appeared to be fine sago. 'Semolina', he said, 'I have it every morning for breakfast'. On the top floor, from which there is a bird's eye view of Temora, the final product was stored. Mr Pardey produced a steel tester which he ran over the top of the flour, drawing attention to the fine smooth surface. 'Pardey's best', he said proudly.

When it opened on 10 January 1908 the mill was capable of producing one ton of flour per hour with the flour 'Snow Queen' being of 'excellent quality'. Town and Country magazine reported the trial lot of bread produced as being 'of beautiful colour, and excellent quality'. It also reported that the 'mill was second to none' and the district was urged to support the firm who had spared no expense in providing a business. The flour mill became the main secondary industry of the town, drawing most of its raw materials and staff from the surrounding district. Staff employed included a foreman miller, shift millers, topmen, packermen, chemist, storehands, head storeman, bag handlers, travellers, carters, cleaners, bag branders, export flour supervisor, depot managers and office staff. Worker interactions with wheat and flour in the mill were very embodied, accidents reported in a mill diary including pulled muscles from lifting a 150 pound sack from the scales to the sewing machine. The physicality involved in lifting wheat in this way is encapsulated in Figure 6.5.

Figure 6.5 Wheat lumper – Temora NSW, c1925
Source: Mitchell Library, State Library of NSW, bcp_02897.

Wheat for the mill came in by rail and road, by horse and bullock wagon from many places, including Ariah Park, 32 km to the west, where today a somewhat ramshackle wooden freight wagon in the main street commemorates the first bulk grain haulage in Australia, in 1916, and the shift away from bagged wheat (Figure 6.6). This momentous event in the transformation of wheat landscapes brought Ariah Park to national prominence but eventually sucked the life out of the town. Like other points of friction across the rail network, trains no longer stop at its station.

Figure 6.6 Experimental wheat truck memorial – Ariah Park, NSW, 2007

Pardey & Co. ensured that top quality grain was purchased by awarding substantial prizes for the best crop of wheat. Each harvest season, a new hat was awarded to the owner of the first load of grain delivered to the mill. It was related that often the grain would be bought in 'too green' in an endeavour to be the first.

A number of changes and upgrades occurred over the years, with Arthur Pardey's commitment to an up-to-date plant being well documented. Many of these changes were to enhance the smooth movement of wheat in and out of the mill. An automatic weighbridge was installed on the company's private railway siding in 1955, where trucks of bulk wheat were weighed, unloaded, reweighed and the tare weight calculated. The mill worked around the clock with three shifts of millers and hands. Plans were made to produce starch reduced flour and hi-ratio flour for use in cake making. The flour mill at this time was the largest customer of the local railway, as well as the biggest user of power in the Temora district. Local industries such as poultry and pig production were able to buy locally made bran, pollard and stockmeal.

Pardey's flour went out to Sydney, Colombo, Penang, the Dutch East Indies, Kuala Lumpur, Seremban, Palambang, Batavia, Macassar, Tarakan, to most areas of the Pacific, China and Russia, but export trade eventually declined as many Pacific countries developed their own mills. Wheat was then exported as grain instead of flour. Pardey & Co. established bakeries to ensure a market for their flour, and the mill continued to operate long after many independent millers were closing. As the demand for export flour decreased and competition increased, production was cut back to a single daytime shift, with occasional orders keeping the mill working into the night.

Les Pardey (Arthur's son) apparently shared his father's commitment to modernity and progress, having intended to diversify into feedlots shortly before his death in September 1973. Two months later, on 27 November, the mill closed down. All the shares were sold to Allied Mills who closed the mill down permanently and destroyed all the machinery in the mill so it couldn't be used again, presumably by competitors. It is believed that the machinery is buried in the grounds near the flour mill. The building was used by Allied Mills as a depot for storage of wheat. It is now the property of BFB (grain merchants, fertiliser dealers, fuel distributors and general carriers), who purchased the mill and silos from Allied Mills in 1986.

The mill remains vacant but in good condition and has Local Government listing on the State Heritage Register. The property is fenced off to deter vandals. The silos are used to house grain. A reconstruction of Pardey's flour mill has been built at the Temora Rural Museum, housing the Ruston & Hornsby Mk 10HRC oil engine used in the mill until its closure. The motor turned over again on Open Day at the museum on 8 March 1986 'to cap a remarkable restoration job and vindicate the time and effort of the museum's small band of enthusiasts'.

From Bags to Bulk, and Back to Boxes

It is easy to imagine the development of the rail network, and the related shift from bag to bulk transport, as an inexorable increase in flow, with sources of friction and punctuation suddenly removed. But, like other transitions in agricultural history, these have had difficulties and friction points of their own, as detailed by Robinson (1976), Whitwell et al. (1991) and Kronos (2002). The transition to bulk handling, like other changes on early twentieth-century farms, had implications for the lives of horses and bullocks as well as people (Figures 6.7–6.9). Although bulk handling had been established much earlier in both the USA and Canada, there was resistance in Australia due to high capital costs, and difficulties in estimating appropriate sizes of the country silos given large variations in harvest size from year to year (Whitwell et al. 1991: 96). Other necessary innovations included conversion of railway carriages, suitable mobile elevators to operate between farm and silo, and mechanical shovels. There were also problems with the lack of depth in Australian ports, ships often having to re-berth at a second stopover port to top up loads. NSW was the first state to take up bulk handling, passing the 'Grain Elevator Act', which saw construction of terminal elevators in Sydney and Newcastle, and country storage facilities, in 1917 (Whitwell et al. 1991: 96). However grower participation was not compulsory until 1934, and even then was controversial.

**Figure 6.7 Loading wheat at Sunshine farm. Horse on left is pulling base of
loader to raise bags – Temora, NSW, c1925**
Source: Mitchell Library, State Library of NSW, bcp_02840.

Figure 6.8 A wagon load of wheat – 18 bullock team – Temora area, NSW, c1919
Source: Mitchell Library, State Library of NSW, bcp_02968.

Figure 6.9 Wheat wagons and the first double wheel truck to bring wheat to Ariah Park Silos. Everyone came out to see the truck – Ariah Park, NSW, c1936
Source: Mitchell Library, State Library of NSW, bcp_02920.

The two World Wars influenced the process, particularly via the supply of jute. The planty qualities of jute (two species of *Corchorus*) produce fibre and textiles useful for wrapping other plants, particularly as agricultural commodities like wheat or cotton. But in Australia, jute bags were expensive to buy and maintain, and grain was increasingly subject to mice and weevil infestations, as well as spoilage from weather. It was estimated that a quarter of the crop value was lost on bagging (Whitwell et al. 1991). Jute supplies from India were disrupted during WWI, when the plant was diverted to the war effort (in the process making European capitalists in India very wealthy). The Indian jute mills 'supplied 1.4 billion sand bags, 713 million yards of cloth, 5 million yards of canvas and a million pounds of twine to Allied governments on war orders alone' (Goswami 1982: 148).

During WWII the Australian government controlled the supply and distribution of bags to ensure minimal disruption, but after the war supplies of jute were severely disrupted by Indian partition. Whitwell et al. argued that 'the Jute factor' had ended argument over costs and handling of bagged wheat. War also created both opportunity and a range of problems for the development of the bulk handling program, including lack of labour, ships and capital.

The establishment of what could have been a single flow of grain depended on a basic standard. The method which became known as the FAQ (Fair Average Quality) system had begun in South Australia in 1888, as overseas trade developed in the latter part of the nineteenth century, and was subsequently adopted in other states. The sample on which the standard was based reflected the bushel weight of the South Australian crop for that season (bushel weight being a measure of density and a good proxy for milling quality) (PIRSA 2011).

Gradually however, grain segregation became an increasing issue. The post WWII breeding programs and expansion into the higher fertility soils of northern NSW led to the growth of premium wheats, for which growers expected premium prices. But of course this wheat needed to both be identifiable and to be kept separate from lesser quality wheat. In 1960–61 a modified FAQ system of limited segregation was trialled; protein content was tested in NSW, Queensland and South Australia, but not in WA and Victoria. Limited storage capacity and bumper crops through the mid-1960s made segregation rather impractical. Handling facilities had to introduce quotas, and farmers were forced to expand their on-farm storage. A global wheat glut eventually forced more market segregation and quality separation in Australia, and 'The Wheat Industry Stabilisation Act 1974' formally ended the FAQ (Whitwell et al. 1991: 109).

In his 1997 Farrer Memorial Oration, Mr Trevor Flugge explains why this shift was so important to a passionate wheat marketer. To most Australians Flugge is better known for his role in running cash into Baghdad during the Australian Wheat Board Iraqi wheat scandal (Overington 2007). In his oration to honour wheat breeder Farrer, Flugge located the wheat industry within a familiar rhetoric of nation-building, overlaid with themes of innovation, flexibility and responsiveness to markets. Under the FAQ system, he argued,

> Australia essentially had an undifferentiated product .. 'wheat was wheat was wheat' ... In fact we had very little idea of where our wheat went, what it was used for and how it rated with our competition. Nor did we have any idea what type of wheat the marketplace wished us to grow ... (Flugge 1997: 3)

As a result of the evolution of the payment for quality scheme, driven by Flugge and colleagues,

> There is now an Australian brand of wheat for virtually every wheat flour use. There is an Australian brand of wheat flour for noodles, steam buns, Middle Eastern flat breads, loaf breads, cakes, snack foods and pastries. Within the brands more than fifty different wheat products are offered to customers, each targeted for specific wheat flour based end products. The Australian Wheat Board now markets to more than forty different countries but our marketing effort is tailored to meeting the needs and requirements of Australia's one hundred and twenty six individual customers. (Flugge 1997: 4)

Increasing product differentiation intersected with varying modes of mobility, providing new sources of friction between road and rail transport. Rail is the preference for export wheat, because the networks efficiently drain 'catchments' into ports. However, rail is a high cost enterprise, particularly in NSW, where wheat has on average to travel further to urban consumers and coastal ports than in other states (500 km versus 350 km national average), and the physical infrastructure is deficient for the volumes moving through the system. Wheat goods freighted on rail struggle to compete with the less seasonal requirements of coal. Companies moving coal are able to schedule in their requirements well in advance, essentially leaving grain companies to take up what is left over after coal has been accounted for. This may or may not coincide with the geographical areas from where grain needs to be moved, or with a firm order for grain at the port. In contrast, road rather than rail is the dominant form of domestic transportation for wheat grain products processed or marketed in Australia. This is due to the greater reach of road networks, and the flexibility and speed efficiencies they provide.

A further recent example of mobility differentiation is observable in the relationship between bulk and container freight. High costs and uncertain availability of bulk freight have led to a situation where container rates are cheaper than bulk rates in some grain markets. This has created interesting opportunities for smaller scale players, as grain trader Malcolm explains:

> Vietnam is a classic example. Vietnam buyers, you know there are flour mills and bakeries and stuff all over Vietnam, and they love their bread, the French influence and all that. All these little bakeries they can't buy bulk quantity of wheat so they have been having to buy it from these re-sellers, and the margins have all been huge. But when you sell them in a container they can buy in lot sizes if they like, and generally it is twenty boxes [containers] which is about five hundred tonnes … a lot of these guys are taking the container right up to the back of the factory, opening the doors, straight in and milling the wheat. And so it is actually quite an efficient chain and the guy can get the hard currency together to pay for it because it is a smaller quantity and it just suits the whole system so much better. (Malcolm, grain trader, 2007)

The differentiation of mobility has required a parallel differentiation in storage options. Several factors are in play here. At the time of Whitwell et al.'s book (1991), permanent storage capacity was below production in NSW and Queensland, meaning that these states had to make extensive use of temporary storage such as tented piles at regional receival centres. In the new deregulated marketing environment, on farm storage has become much more common. Dave's durum-mixing facility earlier in the chapter is a large scale example. At the small scale is the silo and the polyethylene 'Silo Bag'. These have become an important generic risk management strategy to control the timing of selling, rather than having to sell everything at harvest when the price is lowest. Farmers talk about increasing the amount and type of on-farm storage, to some extent becoming their own grain traders. This perturbs grain traders, who cannot track the relevant quantities with ease. At a grain trading conference in Singapore, a London based analyst asked us over drinks about the apparent 'stockpiling' of Australian wheat. He had asked the same question earlier of the AWB spokesperson, who explained it in terms of increased grower control in a deregulated environment, leading to on-farm storage. As detailed in Chapter 5, there is a cost for growers who have to spend a lot of time on the phone and internet to juggle both the physical production and the flows of information and paper.

Trucks – 'if we don't deliver it, you don't eat'

If road freight provides flexibility, speed and efficiency, it is only because of the huge amount of paddling underwater done by transport and logistic companies such as the one we will call the Big Red Truck company (BRT). As the company website proclaims, the business is '… more than merely loading, transporting and unloading a range of goods and commodities. It's about putting raw materials into production lines, products into homes, food onto tables and smiles onto faces.'

As General Manager Adam explained to us in their northern Victorian head office in December 2006, BRT focuses not on harvest grain haulage off farm, which is highly concentrated within one or two months of the year. Rather they have a general freight division servicing large supermarket and food manufacturing companies, a bulk fleet carrying grains, meals, feeds and fertilisers servicing stock feed mills and related processing operations, and a tanker fleet which moves liquid freight such as fuel, lubricants, tallow and vegetable oils, the latter comprising about 60 per cent of the business. BRT thus handles wheat in various forms, carting and tipping it as grain from storage silos to mills, as flour out of the mill to other businesses, or as mill-mix (a by-product from the flour mill) to animal feed processors. They also cart the processed animal feed to piggeries or onto other farms. Substantially modified wheat as ethanol is transported via the liquid freight division. Trucking wheat and processed product is an important physical mediation between the highly seasonal activities of farming and grain production, and the consistent production schedules of mills and processing plants.

The logistics side of the business is a constant process of managing the flow of truck movements, and minimising stops or breaks, as they service the Sydney-Melbourne-Adelaide triangle and the east coast of Australia as far north as Bundaberg. A shift to larger trucks, including 25 and 26 metre B-doubles that can carry over 40 tonnes, helps them offer cheaper rates to customers. BRT is working towards what is termed a 24 hour day, with the truck (although hopefully not the truck driver) working 24 hours for five or six days per week. As much as possible, trucks are scheduled into 'loops' so that they are always carrying something, whichever direction they are going. An important aspect of this logistical management is keeping the trucks maintained and cleaned out from one load to the next. Where tankers and tippers are continuously loading and reloading the same freight into and out of the truck, loads can be topped onto previous loads. However particular loads such as mill mix can build up inside tankers, especially in hot or humid conditions, and truck bins and tanks must be kept clean so that new loads are not spoiled or contaminated. Food contamination scenarios must also be avoided, so for example tippers used to deliver pig feed containing meat into the piggeries cannot also be used to carry meat away from the piggery site. Most mills also have inspection processes for deliveries, and contents are regularly inspected to match deliveries with orders. If loads do not match specifications for whatever reason, a process of interrogation of the truck's previous contents must take place. If the load is downgraded or if the contents are rejected, this can be a potentially expensive cost for the carrier to incur and so, according to Adam, maintaining cleanliness is a major consideration in accepting and organising jobs.

As well as helping to mediate seasonal abundance and regular supply, trucks, drivers and the trucking process also mediate between producers, processors and consumers of food. That this is insufficiently valued in the wider society was indicated by the lengths to which Adam went to explain the significance of skilled drivers, and the difficulties in finding them. Drivers must not only negotiate the physical truck but also the complex regulatory frameworks which govern the conditions under which they work, and the receival and delivery arrangements for their loads. For example it is the driver's responsibility to ensure all the correct paper work is presented at every pick up and load delivery. In some cases up to a dozen forms are required; these might cover anything from grain or flour quality parameters, to cleanliness and contamination checks, or weight and loading parameters. Insufficient information or incorrectly presented data may result in loads being rejected and so these administrative tasks can be a very significant part of the exercise. Drivers and companies operate on very strict timetables with many customers having delivery windows as short as 15 minutes, with the possibility of substantial late fees. In Adam's words,

> … there's an ever increasing volume of freight on the road and yet we find one of our hardest and biggest issues is winning good quality drivers consistently. Good quality as in doing the job – being able to drive the equipment safely, not have accidents, turn up on time, not getting into arguments with the customers, either the pick-ups or the receivers, make sure that they haven't damaged the freight in transit … not having accidents, keeping the equipment clean, keeping it safe, letting us know if there's any work to be done on the vehicles, making sure that they get the vehicles in to get maintenance done on them, making sure they do the paperwork correctly … Plus, you know, there [are] lots of enforcement issues for drivers that they have to put up with … So there's not a lot of incentive for people to go and drive a truck and yet, if we don't deliver it, you don't eat … Even the government doesn't recognise that drivers are qualified. They say it's an unskilled task, which is crap … I don't know about you, but when you drive alone with your kids in your car, would you like to have an unskilled driver drive by you in a B-double combination – 65 tonne on board …?

Ships and Ports

The interviewer gave an emphatic 'no, thank you' to Adam's rhetorical question, but the scale of a B-double with tens of tonnes of load is dwarfed at Port Kembla, where the average sized ship loads about 25,000 tonnes of grain, and ships carrying 60–120,000 tonnes have been loaded in the past. Australia's export wheat exits the country through nodes such as the Port Kembla Grain Terminal, or one of Grain Corp's eight other eastern seaboard terminals. The Port Kembla terminal draws on an area of southern NSW between Dubbo and Wagga, with grain north of this exported through Newcastle and grain south of this exported through Melbourne. 2007 had been a particularly bad year for grain growers in southern NSW, resulting in a very small amount of grain being exported through the Port Kembla facility; in fact the terminal was effectively closed for 10 months of the year. In what would normally have been a very busy time of year, General Manager Matthew had time to talk to us about the business of managing a grain export terminal, at that time depressingly empty. Grain arrives on site by train, is stored for a short period of time and then loaded onto a ship. A very small proportion, less than 1 per cent, of grain also arrives on site by road. Onsite storage capacity is approximately 260,000 tonnes in 30 steel bins. Wheat represents about 85 per cent of the grain moved through the terminal, the remainder comprising barley and canola. In an average year about five different grades of wheat, two of barley and one of canola are handled on site. Each individual ship load of grain begins as an order from an export customer to the company's logistics desk in Sydney. (At the time of our interview, export arrangements for wheat came to Grain Corp via AWB, which was the only company in Australia with an export licence for wheat. This has since changed.) An order, also known as a pre-position, includes the vessel size, destination and shipping date. The logistics department then liaises with the licensed exporter, sources the grain, organises the trains and gets the correct tonnage down to the port terminal within the specified time. Most grain pre-positions are made within 15–20 working days of the shipping date, and an average train holds about 2,000 tonnes, so for example a 60,000 tonne shipment of grain requires 30 trains of grain to be mobilised.

The entire terminal is operated with a staff of about 17, including administration, working regular daily hours, and switching to longer shifts when large ships are being loaded. Matthew explained to us that most staff are reasonably multi skilled but there are a few specialist jobs including electricians. The physical infrastructure itself is entirely automated and controlled from a central control room, with sensors tracking the entire process. When a train driver arrives with a load of grain, a ticket system operates to ensure that the correct wagons are unloaded onto grates underneath each carriage into the correct bins. A train with 40 wagons usually takes about 45 minutes to unload, however trains often carry multiple grades and every grade changeover incurs extra unloading time. Once the grain is underground it moves via conveyor belts into a distribution tower, along more belts and then into the storage silos. Grain is treated for insects on site with either methyl bromate or phosphine gas. Methyl bromate is both more expensive and more toxic but faster acting (three days as opposed to the 14 required for phosphine treatment). Although the company tries to limit the use of methyl bromate, Matthew said that when ships needed to be loaded quickly, it must be used or considerable demurrage fees would be incurred.

Matthew was obviously extremely proud of his terminal and its reputation for reliable, efficient service in combination with a deep water harbour and large storage capacity. Part of this efficiency was attributable to good design, with trains moved through a big circle as they unload, negating the need for any disconnections or shunting. In contrast to older ports, Port Kembla also uses conveyors rather than bucket belts which are slower and require significant maintenance. The main limitation on ship sizes loaded at Port Kembla is usually destination port capacities.

Just as trucks have short delivery windows, ships incur significant demurrage fees if they cannot unload, clean their holds, receive quarantine inspections and reload their holds in a set time, around ten days. Very rarely grain goes in the opposite direction – received from vessels and loaded into the terminal. In the particularly bad season of 2007, grain from Mackay in Queensland was unloaded at Port Kembla and then taken by rail to flour mills in western Sydney. Receiving grain at the terminal does not require extra infrastructure but it is expensive because trucks are required to move the grain between the ship and the storage silos.

Flows of Grain and Information

In the course of this project we had the opportunity to observe many flows of wheat, its constituent parts and products made from it (Figure 6.10). It flowed through the hands of farmers and millers as they extolled the quality of their product; one miller talked about his flour flowing like sand or sugar. We could track flows of virtual wheat as currency traded through different markets. It flowed across computer screens in various symbolic ways – on graphs, in recipes, in mixtures of stock feed (Figure 6.11). Some people as they explained these screens to us were very careful to describe them as a representation of what was happening. Others, who spend their working lives in front of screens, spoke as if the jumping graphics *were* the movements of wheat.

Figure 6.10 Physical grain flow at the stock feed mill

Figure 6.11 Virtual grain flow

So, for example, in describing his day, Malcolm the grain trader told of a working life structured around other fungibles connected to but separate from wheat – money, currencies, loans.

> Okay … most of us would fire up the computer when we're at home, have a look at what the market did overnight … So we'll have a look at the market and start thinking while you are on the train or whatever … Start anywhere from seven-thirty to nine o'clock depending on what is happening. Give the obligatory … look at the emails … we've got a range of people that provide us with information about what has gone on in the markets overnight, some brokers and so forth. So try and read some of that … a lot of our day will be a whole range of administrative things, looking at the accounting issues of the business. … We would then probably talk to brokers, so we'll ring a whole range of brokers who broke the physical commodity markets … our focus is back at the farmer, we are quite happy to use brokers because it is an efficient channel to market for us and they are going to speak to twenty buyers and come back with the best bid … So we'll talk to them, we'll develop a view about what we want to do from a selling point of view perhaps that day. So we'll start to think about that. We'll be managing our stocks, so we've got all this stock, all this grain sitting in a whole bunch of warehouses around the place and it needs to be managed. So our figures will say that we own 358.26 tonnes in one location and that won't match what they are saying so we'll have to go and sort that out. We run pools and of course … the issues there need to be worked through. Just thinking about today – we fund the business so our relationships with banks are pretty important. We use the grain that we receive as collateral to pay the farmers. So there is a whole process involved in that and then you need a friendly bank or two who can understand that, and that is quite a specialist field too, commodity finance. Most banks aren't that, they do different things. So it takes a reasonable amount of management. We are obviously talking to farmers all the time and a big part of what we are doing every day is thinking about the products and seeing how we can distribute them better, talking to the actual growers, dissecting the customer base that we've got … What else do we do? Probably try and have some lunch … (Malcolm, grain trader, 2007)

At an international grain trading conference in Singapore, business cards were the nodes of connection, the material transfer of oneself and the important information about oneself. In the windowless ballroom of a luxury hotel, English was the linguistic currency. People did their email on blackberries during more boring presentations, and there was constant movement of people in or out of the room to take calls or network. An Australian hedge trader in a black suit with white spots tells us that being female helps you stand out in this industry. For her, the lectures are a sideshow because industry speakers never say what they really think, and the value of the meeting is to see all her clients in one place. The friction of needing to stand out, and the seamlessness of circulating through the room intertwined.

The first speaker started with a reference to the overnight market. Another gave examples of the 11.3 millisecond response time in the 200 million daily quotes and orders flowing through one mercantile exchange. The cumulative impression is of the inexorable force of something called the 'economy', responding to massive increases in demand for meat and dairy products in the developing world. Questions of food security loom large, enabling multinational biofuel and plant breeding companies to sound as if they are operating in the interests of the human race in their promise of sustainably increased yields. The steel industry is the driver of seaborne freight; grain is small in quantity next to iron ore and coal. There are 7,000 ships in the world, with 3,300 on order. How does the enormity of this connect to a farmer in the central west of NSW, we wonder, as we munch our way through the complimentary mints? But everything is connected in this world: US household debt, projections of Chinese dairy consumption, Australian pension funds looking for commodities, Philippine perspectives on GM.

Food security issues are inherently geographical, and will likely lead to massive changes in specific landscapes in the coming years. For these traders the world is their playground and the shifts they are discussing come down to the shading or arrows on a PowerPoint map of the world. Australia is the most food self-sufficient country, producing 237 per cent of the calories it needs. Japan is the least, with 40 per cent. Japan would have to triple its agricultural land to be 100 per cent self-sufficient in food. Sub-Saharan Africa on the other hand is 'wasting' land, using only 10 per cent of its arable land. This and the Black Sea region are two of the few places where agricultural expansion is possible – the rest will have to increase their yields. Middle Eastern countries such as Saudi Arabia are getting out of wheat, recognising that the water wheat needs is unsustainable in their context. They are looking to invest in guaranteed supply from Sudan, Ethiopia, Turkey and Pakistan.

The graphs and maps are densely packed with information, and it would be easy to miss the micro-scale of analysis at which these people work. To take one example, consider the ways people viewed Australia. The chair had introduced the AWB speaker with the words 'Australia is being eyed off by everyone in the room' in its newly deregulated environment. Round table speakers based on the other side of the world referred to watching Australian moisture levels in April and August. European shipping representatives based in Singapore asked us about the influence of the green lobby on blocking expansion of Australian port infrastructure. And, as we mentioned above, analysts were concerned at the possibility that Australian farmers were 'stockpiling' wheat outside a system which could count and measure it.

We had asked the Melbourne grain trader 15 months before whether his traders needed to know about wheat. His response shows how a fine-grained understanding of seasonality and microgeographies is not only part of farmer environmental knowledges, but extends more widely.

... no, you don't necessarily have to know but it sure helps ... if they understand what the cycle is on a farm then that information flows up, they don't have to think about it. They know if there is rain, they know what the impact of that is ... Frost is a good one because in every big frost everybody says disaster, and on a local level it can be quite disastrous of course. But rarely is a frost problem as big as the market tries to make it out to be, and so we always say you sell into a frost ... just because geography dictates that unless somewhere is dead flat, the stuff at the bottom of the gully is destroyed and the stuff at the top is not ... So it is just not as bad as it can be made out, and it tends to be quite localised. It can be different but generally not ... (Malcolm, grain trader, 2007)

The point is that the virtual movements of wheat – indeed the wider 'economy' itself – are constituted by material everyday practices such as calculating, conferencing and talking, mediated by telephones, computers and a dizzying array of statistical tools. And, just as Andy the farmer or Fred the scientist (Chapter 4) connect between the paddock and the Chicago wheat price, so the virtual movements have a clear connection to on-ground practices in the paddock.

Even so, by the end of the conference it felt as if the plants were everywhere and nowhere. We turn in the next chapter to food, to bring them a bit closer to human bodies.

Chapter 7
Wheat Becomes Quality Food:
Bread, Pasta and More

One of the main places Australians expect to find wheat is on the nation's table, where the 'core complex-carbohydrate' can be the most celebrated and fundamental item (Bishop 1991: 32). In Australia, consumption of grain for food has decreased since the Second World War (ABS 2000), but the most recent national survey on the topic found that adult Australians each consume about 11.1 kg of pasta, 4.7 kg of sweet and savoury biscuits, and 20.5 kg of cakes, pastries, buns, muffins, and scones per year (ABS 1999). Over 90 per cent of Australians consumed cereals and cereal-based products (bread, breakfast cereal, pasta, but also including rice) and 67–81 per cent of Australians consumed cereal-based products and dishes (including biscuits, cakes, pastries, pizza, lasagne and hamburgers) (ABS 1999). The manufacture of traditional staples like white bread is still a dominant feature of the Australian baking industry, accounting for more than half (62 per cent) of all bread sales in 2001 (BRI 2003). In many such foods the connection to wheat is both visible and celebrated through pictures of the crop or grain. The connotation of a natural product is also connected to the nation; the breakfast biscuit which presents itself as '97 per cent whole grain', with a stylised head of wheat, is also presented as 'Australia's Favourite Breakfast Cereal. Made by Aussies. Loved by Aussies'. A leading brand of flour encircles the wheat stalk with the text 'Made in Australia since 1898'. Influenced by more than half a century of Italian immigration, pasta has become something of a new staple, being widely seen as cheap, versatile, healthy and child friendly. A pasta manufacturer described it to us as 'recession proof'.

In this chapter we examine some of the ways wheat becomes human food within Australia, using two staples, bread and pasta. Although it is often thought of as a staple crop, wheat is not easily food until it becomes something else with the addition of labour and/or capital – flour, bread, pasta. The work of Roe (2006a, 2006b) reminds us that food should not be given ontological status, but rather the process of 'becoming' food involves relations between humans and nonhumans (plants and animals). She uses examples such as fish becoming sushi, someone preparing organic potatoes, and focus groups engaging with carrots. Here, the grinding and processing is part of the becoming food, but it is also dependent on a much wider network that includes the technologies of mobility discussed in Chapter 6. We follow the becoming of food using contrasting and comparative themes; mass and specialist production, handmade and untouched, quality as standards and quality as materiality.

As a net exporter of food, Australia likes to think of itself as contributing to world food security. This is an important issue both within Australia, where an estimated two million people rely on food relief, and in export destinations such as Japan, Egypt and Iraq (each for different reasons). We note also that the food insecure have in recent years included wheat farmers, the majority of whom do not feed themselves with their own produce. Questions of sustainability in food supply need to address these socioeconomic issues, and also those of climate change and peak oil. In most households in the developed world, staple carbohydrates use a tiny proportion of household income and labour because they use a different currency, oil. Neither business as usual nor simplistic localism provides the answers. In this chapter our concern is to explore the everyday miracle of food on the table for millions of Australians. Our study offers a sense of how different actors engage with the embodied, material transformation of wheat into food, and how food connects different places with human bodies, incorporating distant worlds into the self.

Mass and Specialist

Recent agri-food geographies have invested considerable effort in following the networks of alternative food and fair trade movements in an Anglo-American context (Cook et al. 2006). Penker's analysis of two different types of Austrian bread shows that 'local or short food chains cannot simply be equated with ecological quality and environmentally friendly production processes ... a closer look is needed at the specific relations between a particular food chain and the ecological context of its places of production' (Penker 2006: 377). Although not a detailed ecological analysis, this chapter shows that organic or boutique food chains in Australia, at least as regards wheat, are not 'local' in any meaningful sense.

Handmade and Untouched

Geographies of touch have been put forward recently by Crang (2003), Hetherington (2003) and Johnston (2010). Although touching is closely associated with detailed evaluation of quality in some contexts – the 'freshness of the fruit and the quality of the cloth' (Rodaway 1994: 149) – it is often assigned marginal status in western cultures which prioritise the cerebral. In our contrasting examples the bread is both touched a lot, and not touched at all. In each case the question of quality and health is explicitly connected by the baker to touch.

Quality

We do not wish to imply that boutique food is quality and mass is other, although contests over 'quality' often take expression in the dualism of 'taste' versus 'consistency' (Andree et al. 2007). Rather we play on the two senses of quality, as did Atkins in more detail for milk; 'Material has qualities: quality is material' (Atkins 2010: 14). Wheat grain, flour or food *quality* is something that nearly everyone we spoke to wanted to talk about. Farmers, millers, pig feed manufacturers, pasta producers and organic bread bakers all wanted us to know that what they produced themselves, or contributed to as transporters or processors, was a quality product. Quality and what it meant in relation to wheat arose almost spontaneously out of the numerous discussions we had, and it was most often a knowledge and awareness of quality that the people we interviewed wanted consumers to appreciate. Although attributes of quality are important in industrial and other uses of wheat, it is in relation to food that the issue of quality comes to the fore. At this point it is important to note the difference in the use of the term quality in Australia compared with northern hemisphere, particularly European uses:

> In relation to food production in Australia, 'quality' refers to the maintenance of high standards for safety, consistency and traceability, and thus differs significantly from the notion as it is employed in much of the literature from the UK and elsewhere: this often equates 'quality foods' with alternatives to conventional means of production and distribution. (Dibden and Cocklin 2010: 411–12)

Deconstructing Wheat

The floor of Australia's biggest flour mill is so clean and shiny that our shoes slipped as we scuttled to keep up with Hamish, our UK-born Swiss-trained guide. The mill stands between the railway line and the road, a location that poses challenges for twenty-first century expansion options (Figure 7.1). It is like a six storey building, with the wheat falling down a floor or so at each stage of the process. Here the reproductive apparatus of the wheat grain is 'sheared', 'reduced' or 'broken apart' into components that we know as flour, semolina, bran and germ. The flour supplies three streams of customers: domestic, export and the company's own starch processing plant in southern NSW. Before it enters the mill, wheat from various sources is mixed to keep it as uniform for purpose as possible. As the General Manager explained later, consistency is the name of the game. 'We want consistency of supply, we want consistency of production and our customers want consistency of supply, consistency of product'. Impurities that would blemish this consistency are removed at various stages of the process; for example magnets extract any inadvertent fragments of metal.

Figure 7.1 Big Mill at Milltown

Although the size, noise and vibration give an impression of strength and weight being brought to bear on tonnes of flowing wheat, the roller design is in fact maximised to treat each individual grain with precision and gentleness. This is the key to shearing the grain with as much finesse as seen in the botanical layering of Figure 7.2. (Note however that the diagram was taken from an instruction manual for millers, who are most interested in the different layers. Botanists, who are rather more interested in the seed and what is going to become of it once it germinates, would consider this depiction to be upside down.) Australian wheat is very hard and breaks down easily, a reason Hamish likes it so much as a miller. It comes into the mill dry and has to be wetted down, using 2,300 litres of water per hour. (Apparently British wheat in contrast has to be dried out before milling, if it is not to turn into a cotton wool-like consistency.) The wheat is rolled and vibrated through a number of stages, via different shaped grooves in the steel rollers. The aim of the miller is to keep the outside of the grain as whole as possible, leaving a clean bran. So it is gently split and sheared. The white matrix in the middle is going to become the flour. Part of this comes away at the initial stage. But the rest of the milling process is about scraping away as much of the whiteness as possible. The individual wheat grain will go anywhere up to six breaks, with the flour coming out and being sifted off and vacuumed away at every stage. If the flour is not taken out but continually ground and reground through the milling process the gluten is much more likely to be damaged. All of the grain is used, including the last fifth or so that goes into pellets for animal feed.

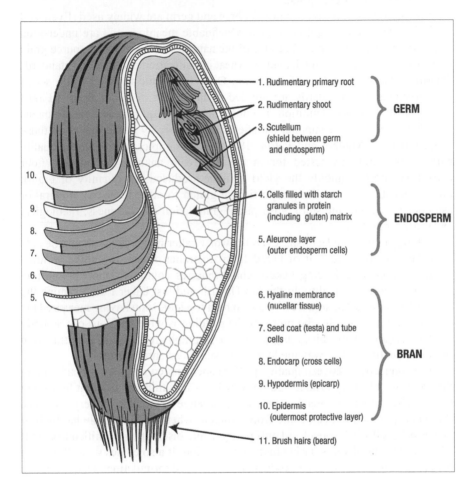

Figure 7.2 Section through the wheat kernel with relevant structures
Source: Adapted from Muehlenchemie (n.d.): Fig. 1, original by Berghoff 1998; also Cornell and Hoveling 1998: 3 and Fennema 1985: 875.

When asked to define flour, Hamish hesitated and resorted to a size definition in microns rather than a substance one. Flour is so ingrained in Hamish's life that he is somewhat taken aback when someone asks him to actually define it. He knows exactly what it is, opening small hatches at various points in the process to grab samples to show us. Using a shiny palette knife that he carried with him, he would skilfully press a flour sample into shape, then rub it between his fingers.

He described the way that Australian flours 'flow nicely'. The interaction between the machine and the flour was a very tactile experience for him, and indicated considerable professional pride.

Although the terms flour, semolina, bran and germ are widely used (Evers and Bechtel 1988), they do not have strictly definable meanings and are understood to be 'variable commodities' because of the natural variations in the source grain (BRI 1989). As we saw in Chapter 6, wheat classification and receival standards determine the end product of the grain. Once inside the mill, a further set of measurements and standards are applied as the wheat becomes flour. '[Flour] quality in its broadest definition, is conformance to requirements: requirements are established and the supplier satisfies the customer by conformance to those characteristics' (Mailhot and Patton 1988: 69). Examples of functional quality parameters commonly tested for in a mill stream analysis include the flour extraction rate (commonly the yield of flour produced per quantity of grain); protein quantity and quality (proportion of gluten to other proteins); mineral or ash content; colour; moisture; absorption; granulation and hardness; rheological and other properties.

Although, to quote Hamish, 'even the worst baker in the world should be able to make a pretty good loaf of bread from Australian flour', the mill's contract to supply flour to a leading bakery chain requires continual laboratory testing to ensure consistent quality. In the lab Hamish and test baker Lulu examine the 'stretch' as they pull apart a freshly baked high top loaf, Lulu having measured the height and width. While 'the proof of the pudding is what the flour can actually produce' (Hamish), machines such as an extensograph give a formal determination of gluten quality. In this way the loaf is 'proved' in scientific terms.

The connection between quality, performance and consistency is headlined in the company's Bakery Product Guide, which outlines thirteen different flours and fifteen different bakery premixes, as well as a range of ancillary bakery products. Product differentiation can come from different wheats, for example hard wheat makes bakers flour for bread and puff pastry, while biscuit flour is milled from soft wheat which shrinks less during biscuit production. It also comes when flours are blended and mixed with other products into premixed formulations such as sponge cake or bun mix.

Big Bread

Australia's largest bakery, in suburban Sydney, makes 17,000 loaves of bread per hour on two main production lines. It has two 50-tonne and two 100-tonne automated silos which signal the nearby mill when they need to be refilled. Because the mill is only 3 kilometres from the bakery, smaller delivery trucks are sent frequently, rather than occasional deliveries from larger tankers. The head baker, who showed us around in April 2007, estimated this 300 tonnes of flour stored onsite represents about one and a half days production, assuming everything is going well.

From the silo or bulk delivery area, flour is pumped along pipes into the mixing area. Other ingredients used in smaller quantities – canola oil, yeast and vinegar – are delivered to the receival store and loaded into automated storage. These ingredients are fed across the room and into one of two bread mixing machines, having been automatically weighed and measured according to which recipe is being used at the time. At each large mixer, water is pumped in with the flour and other measured dry ingredients, which are then kneaded into a large quantity of dough. This mix, taking 280 seconds, is enough for about 400 loaves of bread. The same basic white flour is used for all bread products, with wholemeal flour and wholegrain mix being added for those recipes. The bakery also makes other bread rolls, crumpets and pikelets on smaller lines.

Once the bread dough is kneaded, it is dropped into a divider machine, which weighs out the dough for an individual loaf (about 770 g) onto a conveyer belt. About 100 g of moisture is lost through the entire process, so the end weight achieved is between 650 and 660 g per loaf. Underweight balls of dough can be separated off the conveyer belt and remixed back into the main dough mix. Although 100 g seems like an insignificant figure, this moisture loss factor is a critical issue for the business when multiplied by the number of loaves produced per day. Reduction in moisture loss improves the efficiencies and reduces the costs of the whole process, as well as improving the keeping qualities of the bread.

From this point the dough enters an intermediate prover where it is rested for about six and a half minutes in warm and humid conditions. The resting period enables the yeast to become activated, and the dough becomes soft and pliable. As the intermediate resting period finishes and each ball of dough is moulded, it enters the main proving machine. The dough spends about 65 minutes here at 40 degrees (Celsius) and 70 per cent humidity. According to the baker, this point is where the quality of the flour is very important. The right protein level (about 12 per cent) here gives just enough spring or lift to the dough so that it 'jumps' up in the tin as it is baked. If there is not enough spring in the dough, more yeast must be added at more cost to the whole process. Too much spring and the dough will overflow out of the tin and out of control. After proving, the dough is cut into four pieces and dropped into Teflon baking tins. Square shaped loaves are lidded and high top loaves baked in unlidded tins. This cutting step aids fuller expansion in the tin, stops any holes forming in the loaf and, together with the lidding, keeps the structure of the loaf nice and tight. The lack of air holes also prevents deterioration of the loaf, improving its keeping qualities.

Once the proven dough is 'tinned up', it enters one of two parallel ovens, each of which is about 45 metres in length. These are gas ovens, each heated by 93 burners, enabling the baker to have exact control along the whole length. The loaves take about 22 minutes to cook. They are then 'depanned' and cooled for 105 minutes in a cooler, which drops the internal temperature of each loaf from about 96 degrees down to 30 degrees. At this temperature the loaves can be sliced.

Each loaf is bagged and locked with the 'quick lock' or plastic clip that contains the information about when and where it was baked. The bagged loaves are passed through a metal detector, a standard requirement for all food manufacturing. Loaves which are rejected after baking are either sent to a crumbing machine, and packaged as readymade bread crumbs or, if totally unusable, sold as pig feed. Bagged loaves are packed by hand into coded plastic bread crates which are stacked and then wheeled over to the loading bay for eventual delivery to market. From start to finish the whole process takes about 3 hours and 20 minutes for an individual loaf.

The bread that is made is delivered overnight to the larger supermarket warehouses and sold the next day into the supermarket, or delivered directly to other customers. At the time of interview the bakery had enough demand to bake continuously for 18 hours a day, with about 6–8 hours of maintenance conducted during the night. They bake seven days a week, every day of the year except Christmas and Good Friday, and half a day on Anzac Day.

The entire bread process up until the stacking into crates is automated, and the baker is keen to point out to us that they have excellent hygiene levels, '… you'll notice no-one really touches the product either. You don't touch it in the mixing, you don't really touch it'. About 18 people are required to monitor the two bread lines when they are fully operational. The few staff members stationed on the bakery floor were kept very busy with the constant through-put of product along the conveyer belt, especially where baked and cooled loaves and buns were 'depanned' and checked for consistency. The pancake and pikelet lines are somewhat more labour intensive. The process for these products is very quick, some 8 minutes from start to finish and because these products are softer and more fragile than a loaf of bread, every individual pikelet has to be collected by (gloved) hand and placed into a bag.

Because Big Bread produces so many different products, its bakeries in different cities have partly tended to specialise in particular lines. Most of them produce common products like sliced white bread, but specialty products requiring dedicated machinery are localised – fruit bread in Brisbane, crumpet slices in Canberra. The Sydney loading area also acts then as an intermediate warehouse for Big Bread products from these other places before they are distributed across the state. In the loading bay, the company has recently installed a 'pick to light' system, a series of colour and number coded electronic lights overhanging the bays where stacked creates of bread and other product sit. This reduces the need for labour at the loading bay and speeds up the throughput of customers or delivery personnel receiving goods for distribution.

Small Bread

In a small town on the NSW South Coast, Johan bakes between 400 and 600 loaves of sourdough bread each day, depending on demand. Johan estimates he might be able to produce up to 1,000 loaves. Each week, the bakery receives a tonne of organic stone ground flour from a milling company in Gunnedah, over 600 km to the north. The flour comes in 48 kg stiffened paper bags on pallets wrapped in plastic, which arrive by truck from distributors in Sydney. There are no written contracts for the supply of Johan's flour; delivery is based on a personal relationship with the company which he has developed since the 1990s.

Valerie the organic miller accumulates her year's grain during harvest in Gunnedah, around November and December. Although she and her family have made a large investment in storage capacity onsite, they are unable as yet to hold an entire year's supply of grain, so arrangements must be made with organic growers who can hold their grain until it is required at the mill. Carrying unneeded grain on site can be a logistical burden if the budgets are not calculated carefully. Valerie told us that they had been doing business with some growers since the mill was built and even though she had never met them there were no written contracts for the arrangements that were made. This was a family business, she told us, and everything was built on trust. Valerie purchases grain from a very large area extending from central Queensland through New South Wales, into Victoria and South Australia. When necessary she has purchased organic grain from as far away as Western Australia, and for some smaller packaged products, imported wheat has occasionally been used. Valerie explained to us that sourcing grain from different areas acted as an insurance policy against seasonal variability and quality fluctuations in grain availability.

Grain is trucked into the mill at Gunnedah by a carter who has worked with Valerie's family for years. Organic certification requires that all storage containers including haulage trucks are cleaned and inspected regularly. Bringing grain down from the north into Gunnedah is quite complicated because of road weight limits south of Narrabri. B-doubles must unhitch bins and bring them into the mill one at a time, incurring multiple trips back and forth between the two towns. After the truck arrives, the wheat is augered into holding silos before being air conveyed to a stone retriever and cleaned of any rubbish. Once it passes through this cleaning machinery it is again air conveyed into a milling silo.

Health is a key issue for Valerie who started the business because she suffered from food allergies. Originally she and her husband owned a farm in the district, but with a few bad seasons and high interest rates they decided they couldn't keep 'running the risk'. Her dream for the mill business developed 'out of the blue' from her kitchen baking and reading on the health benefits of whole wheat flour. She attributes the growth in her business to the acceptance of organic produce in Australia, and particularly in New South Wales, over the last 10 years, and also to the recent proliferation of small café bakeries. These customers are valuing health as a higher priority and are willing to pay the higher costs of premium organic produce. She said she feels people are getting better nutrition from the bread made with her flour because the food produced was more satisfying. Valerie reiterated to us that all of their business relationships were built on trust, and that often no written contracts were made for supply arrangements. For example during the previous year when drought had drastically curtailed supply the longstanding relationships, such as those with Johan's bakery, had been given preference.

Johan makes a yeast culture of fermented flour and water to give the bread desirable flavours that commercial – or as he calls them 'ferrari' – yeasts do not. This culture together with the flour, purified water and sea salt are mixed together in a French-designed Italian-made mixer which 'gently' works the dough, replicating hand action, without producing any heat as large commercial mixers might. The dough is then rested and shaped by hand, then left again to rest in individual hand woven willow baskets, originally sourced from Germany, but now from China due to cost.

The 'relaxed' shaped dough is then scored to improve the oven 'bloom' (a description of the rested dough standing up in the oven heat), and baked 60 loaves at a time in the wood fired oven. This is a critical piece of infrastructure for the business, which Johan spent a good deal of time researching. Originally taking months to build, Johan chose the wood fired oven based on a daily baking schedule, as opposed to a continuous wood baking system. According to him, this is a much more efficient use of wood. Johan sources hardwood timber for the oven from a local firewood merchant, and it arrives at the premises on pallets wrapped in plastic. The quality of the wood greatly affects the length of time it takes for the oven to both heat up and cool down.

More water is required in Johan's process than in Big Bread. Between 7 and 8 litres of water are required for every 10 kilograms of flour, a hydration of 70–80 per cent. This compares with commercial bread making which might run at about 55–60 per cent hydration in ideal conditions. Hydration is dependent on a number of factors; for example drought affected flour necessitates higher hydration. Specialty flours such as spelt also require adjustments. The increased moisture in the dough enables Johan's bread to cope with the extremely hot temperatures of the wood fired oven, blooming or expanding and achieving the desirable crumb texture. According to Johan, the increased moisture also increases his bread's keeping qualities.

Most of the loaves are taken from the production bakery and sold on the nearby café site. About 30 per cent are sold by distributers in markets and shops as far away as Wollongong and Canberra. Johan describes his clientele as varied, ranging from 'hippies and alternatives, lawyers, affluent locals and families with kids'. Matching the cost of the product with turnover is a very important consideration. Johan favours higher turnover with lower margins rather than pricing his product more expensively and it not selling. However, care and slowness are also associated with quality, as expressed in the need to keep the wheat 'happy':

> I guess in a way then once it goes in my mixer, the way that I mix the dough, and the way that I do my production, just enhances it and looks after the wheat. I guess it's about slow mixing which doesn't damage the wheat and keeps the wheat happy and healthy. (Johan)

The handcrafting is embedded in Johan's philosophy of food production, which includes respecting the work of the organic farmer and the wholegrain organic miller up the chain who have also treated the wheat grain with appropriate care; they 'look after it' and treat it 'gently'. This philosophy extends to the distributors, as Johan checks out their credentials to ensure they are like minded and share a similar 'organic' philosophy.

> ... what I do is sort of slow and gentle. So you sort of keep the wheat, be nice to the wheat. I know how hard that farmer would be working to get his product to me. So I really feel like I should respect that, and respect him for working so hard.

Touching and Nature

The most obvious difference between the production processes of these two loaves of bread is that for Big Bread, quality is associated with not being touched by human hands, whereas for the Small Bread, the 'handmade' label is literally associated with physical touching, and the dough is handled with ungloved hands at several stages in the process. Johan made this comparison explicit when telling us about a friend who worked in a large commercial bakery: 'I remember Matt told me a story of someone who went and got a job at Big Bread or some bakery like that, and he went to touch the dough and they said, what are you doing? You're not allowed to touch it.'

For Johan, handling the dough is part of the checking and monitoring process. Thermometers are used but, when questioned about the science and technology involved, Johan replied that essentially he is 'feeling' his way through. In his view he was constantly managing variability – of seasonal variation in flour, of wood and its effect on oven temperature, of humidity – whereas conventional bakeries take all the variability out of the process (Figure 7.3). The Big Bread baker on the other hand is proud of his capacity to deliver a totally consistent product. To do this he also has to manage variability, but does it by monitoring systems built into the baking machinery.

Johan positions his bread as a traditional, 'natural' product, made of 'all products from the earth ... a natural handmade product'. Here 'the material operates as a sign for the natural' (Kearnes 2003: 144). Paradoxically, this view of nature elides the human labour that has produced it, including transporting the grain and later flour from distant places. Johan's explicit positioning is also in opposition to bread produced in a commercial bakery, which he describes as 'appalling'.

> It gives you nothing. It might be half the price of my bread but it doesn't even give you half of what my bread does. So you can have one slice of my bread and be satisfied or you can have a whole loaf of that and not be satisfied ... they just basically destroy the wheat, it's just lifeless insipid bread. It shouldn't even be classed as bread really ... It has no body, it has nothing. But obviously it has a place in the market ...

Figure 7.3 Making sourdough at Small Bread

Big Bread baker was less keen to present his bread as 'natural', although it was implied in his noting for us how few ingredients went into the dough mixer. The ingredient declaration on Big Bread's basic white loaf includes 'Wheat Flour, Water, Baker's Yeast, Vinegar, Iodised Salt, Canola Oil, Wheat Gluten, Soy Flour, Emulsifiers (481, 472e, 471) and Vitamins (Thiamin, Folate)'. Big Bread advertising is less reticent, claiming to be made with natural ingredients and have 'no artificial preservatives, no artificial colours or flavours'. Big and Small Bread are both rhetorically deploying nature. In this sense they are both right; there is a demonstrable connection to the products of the earth, albeit via different chains of mobility. Neither is a local product when looking upstream, and both are relatively localised in the downstream direction.

Big Pasta

There are only two or three industrial-scale pasta manufacturers in Australia but they compete with imported product. Italian manufacturers in particular pay substantially reduced labour costs, meaning that, despite higher transport costs, imported pasta is often similarly priced for consumers. (This is a more significant issue for pasta production than bread because of pasta's longer shelf life.) The one we will call Big Pasta make a range of pasta, noodle and other wheat based products in a large factory in suburban Melbourne. They employ 64 people across three daily shifts. Compared to bread, making pasta is a relatively simple process. Semolina[1] flour and hot water are continuously fed into a mixer at a constant rate and mixed to form dough. This dough is then squeezed under pressure through a variety of metal moulds which give each pasta type its shape. The 'wet' pasta is then dried enough to be packaged.

Both durum and non-durum (primarily Australian hard wheat) semolina are used at Big Pasta. The balance between them is directly influenced by the cost of transport, in the process challenging the mystique of durum wheat as essential for pasta. The durum semolina comes from Australia's two largest semolina millers in Brisbane and Gunnedah. Semolina is accumulated for the year around November, after the harvest, with millers presenting Big Pasta with their specifications and prices for the coming 12 months. The Melbourne site has storage for about 700 tonnes of raw material, sufficient for a little over a day of production, so deliveries are made by road tanker on a daily basis. This 'just in time' production process is common to city-based manufacturers for whom the costs of storage would be prohibitive. But carting milled semolina over these distances is expensive and requires specialised haulage infrastructure.

On the other hand, hard wheat flour can be sourced locally from southern NSW and Victoria. The production of pasta using non-durum wheat is a response to this particular market challenge. Big Pasta now have two ranges, one we'll call Big Premium, sold as 100 per cent durum wheat product, and the other we'll call Big Basic, sold as hard wheat product. Commercial Director Brian explained that they had found it difficult to produce their basic pasta lines in a cost effective manner with 100 per cent durum wheat, which was sometimes as much as double the price of Australian hard wheat.

> … you don't make a lot of money, it's a volume business, so hence we've looked more and more on wheat flour and how can we improve the wheat flour that we've got; both that ingredient and then maybe improve it with other improvers to get it to the stage where it's fit for the purpose that it's required …

> … Whenever you go to someone's house for dinner you'll say, oh here's a nice bottle of wine… you're never going to pull out of the cupboard and show them what the pasta is. So it can be a house brand pasta, it can be a $2.50 retail packet of pasta but it's a good fit for purpose meal … well, what's durum mean? I dunno … (Brian, Big Pasta)

Big Pasta see themselves as challenging the industry perception (and mystique) that pasta must be made from 100 per cent durum wheat, albeit at the budget

---

1   Semolina is coarsely ground endosperm, after removal of bran, and before further reduction to flour.

end of the market. By producing pasta from Australian hard wheat they are still meeting the specifications of their customers.

The production lines are completely computer controlled via a control room which observes the entire plant. Each production line can be set up and monitored from this control room, in much the same way as those systems observed at the larger flour mills and stock feed mills (Figure 6.11). Order specifications and the recipes selected determine delivery of ingredients into the mixer, in which the wheat semolina is mixed with water for about 10 minutes. After mixing, the dough goes through the kneader into a screw in a barrel where it is extruded under pressure through a die or mould. There are eight lines on the Melbourne site, each producing dried, shelf-stable product. The lines producing high demand products such as spaghetti and lasagne run continuously, producing some two tonnes of pasta per hour. Other pasta shapes are produced on separate production lines in batches. The extrusion dies which give each pasta its unique shape are periodically changed according to what has been ordered. All of the equipment, including the Teflon coated bronze aluminium dies, is imported from Italy.

The pasta is dried at between 70 and 100 degrees Celsius in an automated process which takes about two hours, depending on the shape being dried. This drying process is also humidity controlled to ensure the pasta does not fracture before it reaches a stable moisture content, generally about 12–12.5 per cent moisture. Pasta is then put into storage, which involves upright silos in tiered sections that feed into the packaging area. Robots assist with the packaging of the dried pasta.

Other production lines in the plant are dedicated to non-pasta or soft wheaten noodle products, such as those which are added to instant or tinned soup products. On these lines sheets of wheaten dough pass through a cutter which stamps out the noodle. There are also other production lines in the factory producing what is known as a pelleted product, a cereal extrusion (also referred to by the technical manager as a 'lower shear, less fully cooked cereal fix'). These pellets are lightly dried and sold to other manufacturers who form them into their final shape through frying or air puffing. These pellets form the basis of various breakfast cereal and snack food products.

The plant also has an industrial scale test laboratory where pasta and other wheaten products can be made on a small scale. This is important not only for testing regular products but also developing new food applications. One such food innovation has been the introduction of alginates into wheat flour products in a variety of food applications including pasta. Alginates give a very different texture to wheaten products and improve the bite or mouth feel of various products. According to Brian, these developments are essential in a market place where manufacturers are always being squeezed on dollars; 'If you can make something that's perceived to be very good at a much lower price we can do OK, and everybody else is happy also'.

The primary export market for Big Pasta is New Zealand, with all the packaging for that country done on the Melbourne site; their New Zealand trading partner shut down their own plant to achieve scale efficiencies between the two businesses. The company sells some pasta pre-packaged into Asia and the South Pacific; however most of their export business is industrial pelleted product, sold to food manufacturers worldwide including China, South Africa, Brazil, Japan, Hong Kong, Brunei, Malaysia, Singapore, Fiji and the South Pacific. In these cases Big Pasta deals with traders who specialise in selling food products into particular countries. Most of the product sold into export market is sold in bulk containers, with the food manufacturer of the destination country processing it before sale. According to Brian, food manufacturers in these destination countries are particularly seeking Australian product because of the high quality standards and consistent clear certifications provided by Australian manufacturers, particularly since the Chinese Melamine scare.

Small Pasta I

An hour's production at Big Pasta is equivalent to a week's production at Small Pasta I, operated by brother and sister Marco and Theresa behind a small shop front in central Wollongong. They manufacture about one tonne of pasta per week, selling 30–40 per cent more in winter, but the business is steady enough year round to be producing seven days a week. Dried pasta is displayed along one wall of the shop; spaghetti, linguini, fettuccini, pappardelle, penne, spirals. Along the other side a bank of upright fridges and freezers displays four or five varieties of lasagne, meat and vegetarian ravioli, stuffed cannelloni, a dozen or so varieties of premade pasta sauces, and pre-grated premium parmesan cheese. At the back of the shop a doorway leads into Marco's tiny office, containing a small wooden table laden with desktop computer, piles of paper and stacks of manila folders. The printer and fax machine sit upon shelves above the desk. Behind this office is the factory where the pasta is made.

During our visit the office was busy and noisy. Theresa was finishing a batch of cannelloni and lasagne, constantly moving between the factory and the shop front to serve customers whenever the front door bell rang. The heavy din of machinery rang out every time she opened the door behind the office. Marco was attending to a service man repairing the photocopy machine and an old friend and customer who had dropped by to pick up his order and chat.

Marco and his sister have been running this business since 1986. Their experience mirrors the rise of pasta in the Australian context. The family came to Australia in the 1960s from a family farm in Italy. As Marco described it, they used to sell the wheat and mill the flour, '... that's how we started, bit of experience, but [owned] nothing, just imagination'. When they arrived in Australia they initially set up another business in the southern suburbs of Wollongong. In the early 1980s on a trip to Italy, Marco first got the idea of making pasta in Australia. He visited some shops in his home town of Naples and looked at the small 'labs' [pasta factories] behind them. Back in Australia he read food magazines and looked at the growing articles on different foods becoming available. In his words, 'I'd say "there's something coming", you know, and I saw it coming and I started making pasta. I bought the machines ... being Italian, it's [pasta] in the blood'.

The small 'labs' Marco visited in Italy told him that '... there was nothing to it, you just have to learn by yourself ...', so he stayed a few weeks and learnt the trade. But after years of experience and scores of food awards, Marco told us that 'pasta's not easy to make, it's very hard'. In the beginning, sourcing the right wheat was difficult. Only soft flour wheat was available in Australia, and the semolina that could be bought was not the same type of semolina that Marco had used in Italy. He investigated and eventually found a company 'doing it'. He bought it but it was not the best. In his words, over time, it got better and better. Today things have improved. Although he sometimes still has occasional problems with suppliers, he buys his semolina from northern NSW and proudly declares that the quality is the best in the world. A key ingredient of Small Pasta is egg, which enhances both texture and flavour. It adds what Marco describes as a 'nice, nutty, more aromatic, vanilla type of flavour'. But adding eggs is tricky, you've got to know how to do it. 'There's a knowledge, also how to dry it and when we do pasta you've got to be careful the machine isn't hot or not too cold. It's very tricky, you know and you don't want to take too long. That's how it is'.

Most of the pasta is sold in the shop at the front, or via other local delis, restaurants and gourmet food outlets. Marco has a distributor he uses to sell into some of the Sydney markets and they are now also doing a line of Australian animal shapes in red, green and white (the Italian colours) for the airport and export trade. Most of that pasta is re-packaged into distributor packaging without the Small Pasta brand name. In an industry environment where branding and company names are everywhere this seemed surprising to us. Marco was not at all perplexed, 'what do I care ... if she a good customer I make it for her. What I care? People already know who I am ... on the books'.

Although we can hear and smell it in the background we are not permitted to look at the factory floor. Marco has learned the hard way that his experience can be used to benefit his competitors. Although he is proud to talk in person about the competitions he has entered and the prizes he has won, he is also a bit jaded by these experiences, describing for us a number of difficult episodes where people have stolen his recipes or in his estimation mishandled competition results against him. However, 'there is no competition because nobody can do better pasta than me so I laugh all my way to the bank'.

Small Pasta II

Small Pasta II in a country town in northern NSW, is a vertically integrated operation of growing, milling, manufacturing and marketing to navigate issues of market risk and global competition. According to Roger, the factory manager, the family who run the operation were tired of seeing product grown in the region and then sold into the cities, the farmers who bore most of the risks getting very little benefit in return. In 2005 they opened the pasta manufacturing plant, and in 2006 a semolina mill. All the pasta the company makes is sourced from durum wheat grown on the property. They claim full traceability of their product right down to the seed used to grow the grain.

The factory's workforce of 20 or so make about 17 different lines of pasta, occasionally reviewing demand for the 'slow movers'. At the time of our visit (December 2008), Small Pasta II was producing about 1.5 tonnes of pasta per week. The following year they established new contracts to begin producing 12 tonnes per week by July and then 24 tonnes per week in August. Further production growth would require an upgrade of the factory and would likely require negotiating agreements with other growers.

Roger described how they are trying to retain the original flavour of the wheat itself, but also distinguish their product by the vegetable and herb powder flavourings added into the mix. On the day we visited one of the company's signature products, a fettuccini flavoured with parsley, was being made. A key feature of the Small Pasta II product is the use of technology which relies on low temperature drying. This technology is labour intensive but is said to provide better flavour. As the long strips of soft pasta exit the machine they are looped over rods which are stacked into drying racks to enter the drying room (Figure 7.4). Air is fed into the drying room at a constant temperature and sensors detect when the pasta reaches the correct final moisture level. It takes between 24 and 36 hours to get the product down to the required 12–12.5 per cent moisture content. The company claim that, unlike other pastas, there is no need for sauces as their pasta is a 'meal in itself', and can be simply cooked and served.

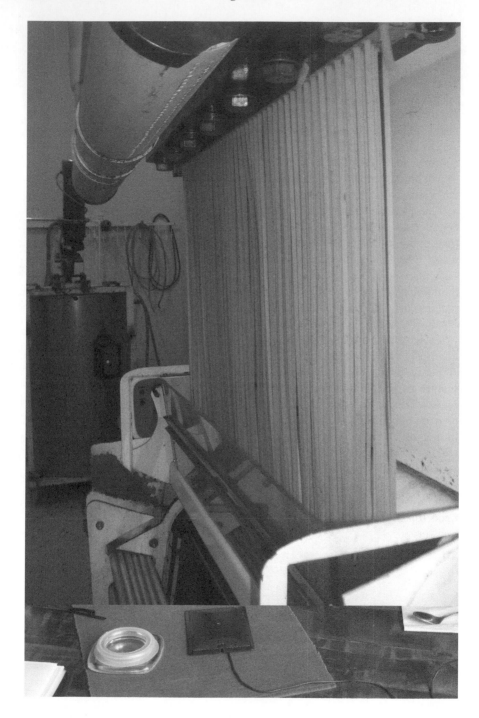

Figure 7.4 Drying fettuccini at Small Pasta II

Long goods such as spaghetti and fettuccini are cut on a cutting machine and trimmed for packaging. Individual bundles are weighed by hand on a set of scales and placed back onto a partitioned conveyor belt and into the packaging machine. Small Pasta II pack into 375 g packets for retail and larger 5 kg lots for wholesale distribution to the restaurant trade. Individual packets of pasta are packed into boxes, stacked onto pallets and moved via forklift into the store room. Most deliveries go by courier as overnight deliveries to the main metropolitan centres. As well as Australian restaurants and supermarkets, Small Pasta II exports to international markets in Japan, America, Dubai, Korea, New Zealand and the UK. The main risk they face is the enormous volatility of the Australian dollar. It can be up to three months between product tenders and final order payments, during which time the dollar value can change substantially. Larger companies can buffer these kinds of fluctuations to some extent, but small businesses do not have the same access to credit or cash flow, making strategic planning very difficult.

Figure 7.5 All sorts of liquorice to sample

Liquorice – Not Staple

A contrasting example to wheat as staple food is provided by wheat as confectionery. We had never thought about flour being a fundamental but unseen ingredient in liquorice, along with molasses, liquorice root, and water, amongst a few other things. In a textbook example of adaptive re-use, Nick makes liquorice and other confectionery in the original Junee flour mill which he bought and re-modelled in 1998. It is an impressive building with cathedral ceilings 20 metres high, and walls up to half a metre thick, and much cooler than the hot outside.

Nick is running an hour late for our appointment, and invites us to sit down for the interview in his café at the front of the building, where he plies us with generous serves of coffee and extra sweet hot chocolate. It is clear that business will not stop even though our sound recorder is rolling, with decisions to be made about potential spelt supplies; '200 acres of spelt – he's started harvesting – don't know if he sold it … can you get his number, thanks … sorry to interrupt'. Nick asks the waitress to bring over his sampling jars of liquorice. The bottles are lined up across the table, and he passes around samples of each liquorice, wheat and spelt, chocolate coated and uncoated, from serving tongs (Figure 7.5).

We have spent all day interviewing farmers, including Nick's father Arthur (see Chapter 5) on his farm nearby. We are tired and struggling to switch into a different conversation here, until the chocolate and caffeine kick in. We all try the various samples now laid out on the table and chew away while Nick talks on. Even to our untrained palates, the spelt liquorice has a different mouth feel – less chewy and more bite. But spelt is different to Nick in a number of important ways. In his words the spelt gluten is more water soluble, resulting in a much more finicky and volatile dough. You 'have to be expert at handling spelt'. Jenny remembers making spelt bread at home and mentions it to Nick who becomes excited at this point and grabs her pen and notebook to make a drawing. It is an illustration of spelt and wheat, a family tree representing spelt as an early archaic variety. He draws on …'wheat is over bred without health in mind; spelt is more wholegrain, more nutritious …'

We have heard this story before and while Nick is busy selecting yet more samples for us to taste, we have a quick discussion about the relationship between wheat and spelt. What seems clear is that for Nick the two must be understood together; his understanding of wheat is informed by his understanding of spelt and vice versa. This is not simply about what he has read or heard but also about how spelt and wheat 'behave' in his liquorice vat and in his business in very different ways. Wheat 'holds up', spelt cannot be over kneaded or it collapses. Wheat is chewy and elastic, spelt falls apart. Wheat is abundant; spelt is expensive and difficult to source. 'Your Eastern Suburbs[2] areas are into spelt, whereas your Westies, they like the taste of [wheat] liquorice and they get liquorice'.

---

2   Of Sydney, an affluent area. 'Westies' refers to people who live in the more working class western suburbs.

Why liquorice, we wondered? At this point Nick gets even more excited, reminding us that we would not have been sitting round a café table enthralled by a bowl of flour, nor would he have been able to sell much to tourists:

> we wanted ... to make a product that can go to the consumer ... And you might even take a kilo [of flour] home with you and sit [it] in the cupboard till it gets weevils and then throw it out because why, you're too busy to cook all those cakes that you fantasised about making. So ... There's no-one else doing it, it's a confectionery, it actually tastes good, it's almost naughty in your mind 'cause it's that good and ... then the third one is that it's immediate use. You buy it, you eat it. There's no preparation, there's none of that carry on ... Yeah. So it's a no brainer really isn't it? It's a great pick off the shelf.

Again, this is a textbook example, of post-productivist agriculture, and of value-adding in rural areas. It also challenges any simple view of organic agriculture as having 'short' chains. On one hand, Nick's father's farm a few kilometres away regularly supplies about 40 per cent of the wheat and spelt used to make the liquorice. This familial local connection is built into the branding emblazoned on the products lining the shelves. On the other hand, variable production due to drought, combined with high demand for organic flour, means that Nick must constantly cultivate relationships with more distant suppliers. Like Valerie the miller, who sometimes purchases her organic grain from as far away as Western Australia, Nick depends on connections with suppliers from South Australia and Victoria as well as in the surrounding district.

Nick keeps talking but it is Friday afternoon and clearly things are closing. He is a good salesman and a generous tour guide. We head out via the shop, each of us carrying an assortment of little brown bags.

Small Shopping, Big Shopping

On a rainy public holiday, Small Foods is absolutely packed with customers and products, all squeezed into a small space. The business includes fresh fruit, vegetable and dry goods for sale (including a small home delivery component); take away foods made on the premises; and a café. Produce is sourced through markets and through organic wholesalers and smaller suppliers via paper and fax orders. The main focus is organic, but this is supplemented by non-organic food when organic is not available due to seasonality or quality problems. The shop has an opening to an adjacent naturopathy business, with clientele flowing between the two.

One of the best-selling products is an artisan sourdough bread from a Small Bread company (not Johan's). Store owner Daniel explains its popularity as an unsliced, crusty, well-formed loaf that tastes really good fresh and comes in a range of different types. Food quality is a really important consideration. 'We use good quality products, my business partner is a real foodie ... particular products [are] highly sought after, for example muffins ... customers are ringing up when muffins are baked.' He is also keen to point out that he has a large customer base seeking wheat-free products, the most commonly stated reason being health considerations. The customer demographic reflects the local beachside diversity; '... some health, some environmental, some just want good food, some are fanatical'.

Daniel attributes their 15 year success to the store being attuned to the food quality expectations of his customers, and to direct and personal relationships between staff and customers, relating a number of anecdotes about the lives of some of his best known clientele. These customers are loyal and they bring their lives, their children, their business meetings, their celebrations into his shop. 'We have a reputation, [we're] part of the local community, we cater to that community'.

Small Foods customers will seek direct information from staff and question them about the products for sale. When a customer had complained recently about the scarcity of certain organic produce, Daniel was able to explain where the produce had come from and that that area was experiencing drought. Apparently the customer was very satisfied that someone had taken the time to explain this problem to them.

On this relatively small scale, a food store can provide more than just quality food; it can help create a sense of community in a very busy place. The staff are in a position to provide some form of education and information to their customers about the food they are buying and eating.

We ourselves encounter the dilemmas of shopping, eating and feeding people – big or small, specialist or mass, cheap or expensive? – in our everyday roles as mothers and household managers. Even when professionally sensitised to the ubiquitous presence of wheat, as we had become, it is not always foremost in our attention when shopping, as the following account, based on our field notes, exemplifies.

***

I've just dropped Ethan at school and need to pick up a few things from the shops before Sophie's appointment in town. I can't do it later because she's too tired and will need lunch and a rest before we pick Ethan up at three. I feel guilty about short car trips to the shops, but it's too far to walk and carry everything with the kids.

I need bananas, breakfast cereal and milk so we'll have to go to Big Supermarket because that's the only place I can get Ethan's lactose free milk. Sophie is hungry before we get there. I've anticipated this. I'm armed with a drink bottle and morning tea from home – a snack in pieces that she can hold onto without dropping it all over the supermarket floor, but that will occupy her for long enough. We always have to start at the fruit and veg because that's where the trolleys are. With her in a trolley I can move faster through the aisles, past all the things we don't need, but starting with fresh stuff annoys me because then everything else sits on top of it.

We pick up bananas and strawberries, an extra thing Sophie insists on which I immediately give into because its fruit and that's OK. Straight through to aisle three and breakfast cereal. First the oats, the muesli, then the women's cereal, the sport cereal, the high fibre cereal, the individual portions, the kids' stuff, is that cereal? Immediately Sophie points out the green crocodile one, the yellow daffy duck one, and the pink Dora cereal. Before she starts up I assert that we don't buy this cereal because these are not healthy food choices. As soon as I do so, another shopper passes me, selects the green crocodile box off the shelf and puts it into her trolley. I'm embarrassed, making a mental note to come up with a different reply.

We make it to the weet-bix or, as Ethan calls, them the boring bix. There are seven varieties; weet-bix with three flavours of reconstituted fruit, extra high-fibre fruit, multigrain, organic and plain weet-bix – all in family, regular and individual portion sizes. We usually just have regular weet-bix. I've read the label at home over breakfast enough times to know what's in it. I figure if you want to dress up cereal, you do it at home with fruit and yoghurt, or honey. But Sophie picks out the yellow box. I hadn't noticed that these boxes were different colours; I've only been reading the labels, not looking at the colours. The box we usually get is blue – 'that's for boys mummy'. She wants the yellow box. She doesn't know what's in it. She doesn't care if it will fill her up and keep her satisfied. She doesn't know what minerals or vitamins are in it, and she certainly doesn't know all the things that are *not* in it. Yellow is the multigrain weet-bix. Yellow is a choice I can live with today ... don't fuss. I'm sure the people who make the weet-bix know about the yellow colour on the box.

Keep going, avoid aisles 6 (biscuits), 7 (lollies) and 8 (chips and soft drink), down to the back of aisle 9 to get the milk, and then back up aisle 9 to avoid the ice-cream at the front of aisle 10. Just the checkout to go and we're out.

Plantiness, Quality and Trust

In this chapter we have traced several examples 'of how a thing (an animal or plant) passes through a set of human practices and material processes that *do* the translating from food production to food consumption' (Roe 2006a: 109). Or rather, we have traced the way the same thing – wheat – becomes different examples of similar foods, bread and pasta. (And we did not actually get to the consumption, apart from Nick's chocolates.) Thinking about wheat as becoming food may be less confronting (or easier to render thinkable) than the blood, skin and death of sushi that Roe discusses, but the bodies of both plants and people were very present in all the processes we looked at. Wheaty plantiness put the flow in Hamish's flour, the stretch in Lulu's high top loaf, the 'jump' in the Big Bread tin, 'held up' in Nick's liquorice and was 'happy' in Johan's bread. Those producers who touched their wheat in an unmediated way provide a clear example of embodied engagement, but this should not obscure two points. First, apparently unmediated touching in Small Bread and Pasta is made possible by just as complex a network as that which underpins Big Bread and Pasta – including for example plastic-wrapped wood for the bread oven and organic wheat from the other side of the continent. The specialist food production examples discussed here are not more 'natural' in any straightforward way. Indeed, partly because they are still minority networks in the Australian food landscape, they seem to be fraught with higher levels of friction than might be suggested by slow, happy or healthy food. Second, the production of Big Bread and Big Pasta is just as embodied, albeit the human bodies are more likely to be reading a computer screen, checking a temperature gauge or driving a forklift. And the wheat is no less planty for being mediated via the aroma of Daniel's muffins, Sophie's yellow weet-bix box, Big Pasta's de-mystified non-durum pasta or Nick's choc-coated spelt liquorice.

Each of these food-makers articulated their own contribution, and their food, through the notion of quality. Quality had a number of different constructions; it was variously healthy, hygienic, natural, fit for purpose, economical, delicious, chewy, consistent, fast and slow. Some of these expressions of quality related to the wheat itself, its material properties, and the careful selection of the right wheat for the task at hand. Others referred more to the total network of technology, labour and product that contributed to a high quality outcome. Among both farmers and other contributors to the food production network such as those in this chapter, there was a palpable sense that people feel they work very hard to produce good quality food. They feel this effort is undervalued by consumers, and underpaid by anyone downstream of their process. 'Consumers have no idea how cheap and plentiful food is' and 'consumers really don't appreciate how complex the whole system is' were common refrains.

If 'trust is a means of coping with complexity' (Atkins 2010: 154), we draw attention to the different ways quality and food can be trusted, or not. On the one hand there is the lack of written contracts between Valerie and Johan. On the other there is a complex of standards and accounting systems for wheat and flour quality, not to mention food hygiene and organic certification. Similar to the transport systems of the previous chapter, both the written and unwritten systems are part of the infrastructure to enhance the movement, processing or transformation of wheat as it becomes food. Each requires different types of human labour, and the differences help us think through what kinds of trustful relationships might be most workable and sustainable in the future.

Underpinning wheat and flour standards is the human attempt to impose some order and uniformity on the mobile and fractious identity of wheat, that variable commodity and category. Wheaty plantiness, in the form of different levels of starch and protein, intersected here with the economics of trying to 'add value' to a raw agricultural plant product by, arguably, taking it further from its planty status. In the next chapter we meet some wheat people whose daily professional challenge is to dismantle that identity further, and reconstruct it in different ways.

Chapter 8
Transformation: Consistency, Craftability, In/visibility

The uniquely craftable qualities of wheat take it far beyond being a staple food, important as that is. It can be pulled apart and reassembled in many different ways. Some of these re-assemblages are edible, but many are not, hence wheat's status in the market as commodity rather than food. In this chapter we explore a number of those transformations, focusing on the material rather than metaphorical, and travelling between the laboratory and the supermarket and back to the farm. Just as other domesticated food plants emerge as 'countless, impossible to avoid' (Cook 2004) and 'invisibly ubiquitous' (Whatmore 2002), wheat is – even more so – simultaneously visible and invisible, obvious and hidden, everywhere and nowhere.

In discussing the food industry, Australia's Department of Agriculture, Fishing and Forestry uses terminology of *minimal*, *substantial* and *elaborate* transformation (Commonwealth of Australia 2008). Wheat exported as cereal is considered minimally transformed. Substantial and elaborate transformations (not specified separately, as the latter are proportionally small) include milled wheat, wheat flour, wheat starch, biscuits, bread and cakes, and pasta. The specifics of those transformations, and their extension into things other than food, are considered in this chapter. In the laboratory and industrial factories, flour produced in the flour mill is pulled apart further, using chemical rather than physical processes. It is fractionated into substances understood as having particular chemical properties, such as the ability to form gels. But these substances are themselves assemblages or categories of constituent things; carbohydrates (as different types of starches and sugars), proteins (as gluten fractions), fats and lipids, fibres, enzymes, and many other compounds (Pomeranz 1988). At the same time as the plant itself becomes less and less recognisable, it is its plantiness which is attracting all this attention. Milled flour becomes: adhesives in box manufacture; additives in biodegradable 'plastic' packaging; source stock for renewable fuels (like ethanol); and alcohols for use in industrial textiles, pharmaceuticals, inks, cleaners, propellants for perfumes, cosmetics and personal care products (www.manildra.com.au).

For most Australians, wheat is so much a part of daily life that we interact with it unconsciously. In contrast, sufferers of coeliac disease have to be highly attuned to its presence and absence, because of the potential danger it represents. Bell and Valentine (1997) have previously discussed the profound limits to the everyday geographies of coeliacs, arguing that they are 'in particular ... very sensitive to and therefore aware of the subtle and often unpublicised ways manufacturers continually alter production processes and the contents of everyday foodstuffs' (Bell and Valentine 1997: 50). Coeliac disease provides a literal wheat shadow: a way to render wheat visible by following people who have to skirt around it. So wheat in turn becomes transformative; our wheat centred culture defines non-wheat as Other.

As we saw in Chapter 3, wrapping the category of 'wheat' around a taxonomically complex set of plants is itself fraught with complexity. Probing the transformation and material reconstitution of wheat rapidly takes us further along the road of challenging its essentialism. People in this chapter understand wheat differently depending on what they want to do with it. For some, we will see that wheat is just wheat, providing a consistent source of calories and energy. In other contexts wheat is definitely not 'just wheat', but a source of very particular proteins or starches, providing consistency of a different type, 'the condition in which matter coheres so as to "stand together" or retain its form'.[1] There is a long history in English of connecting the 'viscous or firm condition' (of food or other matter) with the concept of consistency as 'agreement or harmony between parts of something complex'. Material qualities of flesh, barley, cream and mud, as well as wheat, are threaded eventually into linguistic representations of coherence and stability. The starchy properties of wheat are shown in this chapter, for example, to provide consistency of both types to the marketability of pet food. But wheat is also inconsistent, troubling notions of harmony and settlement. The picture of wheat as a healthy staple food is in stark contrast with its categorisation by Food Standards Australia New Zealand (FSANZ 2009a) as a substance requiring compulsory identification on all food products, and by others (Helfe 2001) as a 'hidden' ingredient in food.

The Food Label – Rendering Wheat Visible

To consider how transformed wheat pervades our daily lives, we return to the Big Supermarket of the previous chapter. Australia has one of the highest concentrations of market share in the retail food sector, with companies Woolworths and Coles controlling as much as 80 per cent (ACCC 1999) of the industry. These companies play significant roles in both horizontal integration, having branched into spheres such as petrol and perishable foods, and vertical integration with investment in the supply chain. Big Supermarket provides a readily available sample of the thousands of consumer products available to Australians. It is a place built into our own weekly lives; we regularly enter supermarkets as consumers and as eating bodies. We started with the basic question, where is the wheat in the supermarket? You do not have to look far. Much of it is obvious in piles of bread, aisles of cereal, banks of pasta and rows of biscuits. The supermarket is the traditional outlet for

---

1   Definitions and etymology from *Oxford English Dictionary Online* (dictionary.oed. com) and *Collins English Dictionary*.

both Big Bread and Big Pasta (Chapter 7), but 'Small' or boutique versions of various foods can increasingly be found here. Golden sheaves of wheat are often prominent on their packaging, making the connection between the farm and the healthy food of the nation. Supermarket chains are anxious that we consider ourselves close to the farm rather than part of the urban masses who, conventional wisdom tells us, are separated from the conditions of production of our food. Woolworths, for example, markets itself as the 'fresh food people', introducing us in different ways to named farmers. During the period of our study, the connection between the farm and the city was further made by appeals in supermarkets for drought stricken farm families.

To find the hidden wheat in this arena of daily life, a key research tool was the food labelling necessary to protect vulnerable people from wheat. Wheat must be identified because for sufferers of coeliac disease and gluten intolerance, supermarket shelves are a minefield of wheat which must be negotiated daily to avoid serious health consequences. Coeliac disease is a permanent dietary intolerance to wheat gluten and other similar proteins of rye, barley and oats (Howdle and Losowsky 1992), resulting in damage to the lining of the small intestine. The condition can manifest as a wide range of symptoms including indigestion pain, autoimmune disease, neurological problems, infertility, osteoporosis and malignant disease (Maki and Collin 1997, Green and Jabri 2003). The condition is irreversible and dietary avoidance must be strict and lifelong, although damage to the small intestine can be repaired when gluten is eliminated (Howdle and Losowsky 1992). Coeliac disease is thought to be relatively common, affecting one in 250 people (Green and Jabri 2003). An alternative strand of thought argues that this allergy and intolerance is not a disease but rather a phenomenon only brought into being because coeliacs are living in a dominantly wheat eating culture.

Product recall notices on the Food Standards Australia New Zealand website (FSANZ 2009b) provide an insight into the world of 'hidden' wheat. These notices detail products recalled due to 'incorrect' labelling, according to the new labelling code, of wheat and wheat derived products. As well as a bread line recalled due to undeclared wheat, other items include various prepared sauces, flavour sachets in processed noodles and confectionery items. Warnings alert consumers to the presence of undeclared wheat and wheat gluten and the potential dangers to the gluten intolerant. Another alert (FSANZ 2009c) lists products which might contain wheat and so should be avoided by those intolerant to it. These lists catalogue an enormous variety of everyday staples as well as relatively new food phenomena such as processed meats, seafood extender, and hydrolysed proteins (see also Shepherd and Gibson 2006: 159–161).

Current guidelines direct that wheat must be specifically noted in a food product's ingredient list, including where it may be the source of another ingredient used, for example starch, unless the food is not required to bear a label (the ingredients must then be declared verbally or in writing). Similarly, if a food additive such as a colouring or thickener is derived from wheat, that source must be declared on the product label. If a label does not disclose that the ingredient was sourced from a listed substance, then 'logically' the food does not contain that substance. The term 'gluten-free' often used on food labels, is problematic for interpreting the presence of wheat in Australian food, as it relates to a 'no detectable gluten' limit rather than a limit of zero gluten (Shepherd and Gibson 2006).

In this context, the food label is not only a research tool but a significant agent in the assemblage that connects contemporary city dwellers to the conditions under which their food is produced. Any supermarket trip will reveal harried parents trying to contain the consumption requests of their kids, pensioners stretching their dollars as far as possible, and rushed one-basket people doing a quick shop on their way to or from somewhere else. With varying degrees of care, most will read the label of regularly purchased products, either at the point of purchase or later in the home. The ingredients list, nutritional information and country of origin are all embedded in Australian food packaging in more enduring form than is the price of the item. The reliability and consistency of food labelling legislation are not perfect, but as a form of accountability they are much stronger than the subliminal message of the sheaves of wheat.

If the connection to food via the label is useful for everyone, it is vital for coeliacs and their families, who have no choice about whether or not to read the labels. Our interviews with some of these people reveal their dependence on the labels to render the wheat visible. We interviewed six people affected by coeliac disease, either because they suffer from it, or because they support another sufferer as family, friend and/or medical personnel. Four of these participants were recruited through contact with the NSW Coeliac Society and two through contacts with other participants in the larger project. They show how the presence of wheat transforms the sociality of their everyday lives. Melinda is not herself a sufferer but has nine family members diagnosed with the disease. Her family has known about the disease and its symptoms for a long time. As well as commonly missed favourites like sliced white bread, sufferers also spoke about missing things like beer because of the effect on their social interactions. Melinda had been initially surprised how much wheat is in food.

> You wouldn't think there would be gluten in ice-cream, would you, or tomato sauce? I mean, we know soy sauce, but all the other things that gluten is in ... Lollies! I never thought lollies. You know you can't have all these different sweets, you can't have that, that's quite amazing, it does surprise you ... (Melinda)

The initial diagnosis period can be comparatively difficult for adults who have already become accustomed to the tastes of particular foods, even if they do make them sick. Food providers are sometimes questioned at length by sufferers to ensure the food is strictly gluten-free, and a number of our research participants recounted difficult experiences where they have trusted the advice given to them, then eaten supposedly gluten-free food, only to end up very sick later on. They also spoke about fashionable food trends to be 'gluten-free' for so-called health reasons. This has confused food providers about safe levels of gluten in food.

Pauline has coeliac disease herself and has family members who are sufferers. As a dietician she advises other coeliacs about how to manage their food. She sees her experiences as being much easier than those of her mother, who had to find wheat free foods in the days before the new food code. Some older coeliacs describe the availability of gluten-free foods today as a 'veritable smorgasbord'. Pauline teaches her patients how to read labels and which brands are 'OK'. Some participants talked about sharing family recipes and finding substitutes for ingredients as a way of extending their food repertoire. Particular difficulty was associated with eating out, either at restaurants or at friends' houses, and especially while travelling. Pauline described eating at home or eating at trusted friends' homes as a strategy to avoid eating out and the potential danger.

> ... it's a good social thing to do, go out and have a meal with friends, so yeah, I guess that has changed a bit ... they're quite good now so they'll always check with me where I want go for dinner which is really nice, but often we'll have meals at each other's houses now too, that makes it easier as well. (Pauline)

Fiona also suffers from coeliac disease. She told us she had been sick all her life, but was not diagnosed until she was in her mid-fifties. Like Pauline, Fiona had difficulty travelling; 'I'd travel anywhere and feel quite sick'. Ironically the only place she did not get sick was at home, a wheat/sheep farm in central NSW she runs with her husband and two grown sons. We had visited to interview Fiona and her husband about their interactions with wheat. We had expected to be discussing drought, fertiliser prices and succession planning, which we did. But as Fiona served gluten-free biscuits with our tea in her farm kitchen, the tangled network of wheat and its shadows became further apparent. Fiona commented on how much easier it now is to find gluten-free foods in the supermarket of the regional town where she shops.

Wheat in the Supermarket

We used online shopping lists to access and research a publicly available list of retail products. Food labels are not provided in virtual supermarkets and so we visited supermarkets to field check product labels for the presence or absence of wheat. Of the 12,034 items listed on the Big Supermarket online shopping lists, 10,235 were available on the shelves to look at. Availability varied according to factors including seasonality (for example Christmas hampers), locality preferences (organic items were not available in some stores), recalled and discontinued items, and an item's suitability for home delivery service. Unknown ingredients and additives were checked against the literature for potential wheat derivation. (Full details of this study and its methods can be found in Atchison et al. 2010.)

Figure 8.1 Bread at the supermarket

Of the 10,235 items surveyed, 6,627 were food and 3,608 were non-food. Of the surveyed items, a significant proportion of food (9 per cent) and a majority of the non-food (66 per cent) items were labelled in such a way that the presence/absence of wheat or wheat based products could not be identified. The majority of food items without labels were fresh fruit, vegetables and fresh meats, which are sold in bins and not packaged, or alternatively refrigerated with only simple packaging. However there were also processed foods including rice and tea, many of which contained multiple ingredients, which were unlabelled.

According to the labels, 20.7 per cent of all surveyed items (1,977 food and 144 non-food) contained wheat. Most of this concentration is in food, with about 29.8 per cent of all surveyed food items containing wheat. Across all food items, wheat was most commonly present in staples including breads, breakfast cereals, and other baked items like muffins and biscuits (Figure 8.1). These are products where the connection to wheat is both visible and celebrated.

Wheat was commonly present in processed foods such as sweets (including chocolate), frozen meals, packet soups, chips, snack bars, baking needs, cake mixes, marinades and savoury crackers. It was found often, but not consistently, in foods which might not be commonly thought of as containing wheat, including vinegars and dressings, gourmet products, mueslis, vitamins, ice-creams and ice blocks. Constituent wheat compounds (starch, gluten or glucose) form part of the ingredients for these foods, but these compounds may also be sourced elsewhere (for example the starch from rice or the glucose from sugar cane). Canned vegetables, cheese, chilled milk and seafood are some of the foods in which wheat would be least expected, but it was present in some of these in highly processed forms such as wheat-derived soy sauce or derived caramel colour. None of the more processed foods drew attention to the wheat connection on the packaging, since to do so would be to render the industrial nature of their production visible.

Food items that did not contain wheat included gluten-free specialty ranges; wine and soft drinks; yoghurts, creams and juices; and meat, fruit and vegetables, all supermarket-defined categories, which contained both processed and unprocessed foods.

Wheat is more difficult to find in non-food items at the supermarket. We might not expect non-food items to be labelled with the same stringency as food, but these are also things that we put onto and into our bodies –notably pharmaceuticals such as pain relief (Figure 8.2). Many non-food products simply have no ingredients listed. Only 4 per cent of non-food items – most commonly pet food, followed by hair care and hair colour products – were identified as containing wheat. These products were the most clearly labelled categories of non-food items but even in these product lines, many items had unclear or inconsistent labelling. A smaller number of other products did identify wheat as an ingredient, including skin care, cosmetics, and shaving products, although many of these had only partial disclosure of their ingredients (for example, 'active ingredients only'). By contrast, vaginal pessaries sold in pharmacies list non-active ingredients and proclaim themselves gluten-free.

Figure 8.2 Hidden wheat at the supermarket

In fact wheat is a very significant ingredient in the manufacture of some non-food products at the supermarket. One example which neatly illustrates the invisible wheat is dry pet food. Although there is an industry code of practice to which many pet food manufacturers adhere (PFIAA 2007), regulations are state-based and somewhat variable. Some of the more expensive brands of pet food provide detailed information about ingredients and nutrient breakdowns, but many brands simply subsume wheat in other grains which are labelled as 'cereals'. A limited number of 'premium' pet food products even promote their 'gluten-free' status.

Two forms of pet food, wet and dry, are popular in Australia. Wet food is primarily composed of meat, with smaller amounts of cereals. Dry food contains more cereal, and is popular because of its convenience; it does not require cold storage and quantities can be carefully controlled as they are dished out. Apart from minor increases during winter, seasonal demand for pet food is fairly constant. This means that manufacturers employ intermediary services to accumulate and secure the grain from growers, then store it to provide a regular and steady supply.

As one manufacturer explained, wheat is critically important in the manufacture of the dry pet food. As well as being the major source of carbohydrate, it is the wheat proteins, the gluten, that allow the finished extruded product to be formed correctly during manufacturing. Manufacturers rely on this textural property to give shape to the processed mixture that forms the kibble. As with the stock feed industry, grain substitution does occur, but in reality this can only be manipulated to a certain extent. In addition to the grain which is processed on site, tertiary manufactured wheat glutens are added to aid this process.

> Wheat is still probably our critical item … the gluten properties within wheat. So the actual chemistry of wheat provides a product which allows us to create a kibble product so it extrudes well. So cereals, wheat is critical. We can, as I said, on the margins probably substitute to some extent, but … without some sort of significant change in technology, I would still see us consuming wheat decades down the track. (Manager, Fast Moving Consumer Goods Company)

The competitive nature of this industry is seen in the company connection this man chose for his own label, 'Fast Moving Consumer Goods Company'. The shape of the extruded product is central to brand identity. Extrusion technology is a closely guarded industry secret and we were not permitted to tour the factory. The plantiness in hidden wheat thus increases the capacity of the company to make its product visually distinctive and more marketable. It helps expensive long-life dog snacks look like bones.

Reassembling Wheat

The craftable qualities of wheat are expounded by food scientists Dennis and Kevin, and starch scientist Jeff, who work in major companies to create new products such as different types of bread, noodles and starch-based products. They both meet and create consumer demand; Dennis talks of 'prospecting' for new products and projects. Wheat is an exciting raw material, and their interaction with it a source of considerable professional pride for each scientist. The complexity of wheat is what their work is all about. So, the wheat 'shows up in just about everything', in Kevin's words, but it can also be hidden, and is not a single coherent entity. 'Wheat isn't wheat. Wheat is lots of different types of grains for different purposes' (Dennis).

Kevin's specialty is making the link between grain quality and food products. His team evaluates new varieties for their potential in the market place. They mill flour in test scale mills, analyse it in a laboratory kitchen and then bake it in a test food laboratory. In one particular test kitchen, the Asian food laboratory, they are able to manufacture and assess products including Asian noodles, Chinese steam breads and Middle-eastern flat breads. Laboratory analysis might ascertain wheat colour and colour stability, brightness and texture. The interplay between new technologies, new markets and new products requires liaison with breeders and growers, who also become part of the transformative system. Each of these scientists is involved in different ways in this system, whether through direct industry liaison or as part of advisory panels and regulatory committees, reflecting the links between the commercialisation of wheat breeding and the development of new products. For example, Dennis is also a member of the Australian Wheat Board Classification Panel, which meets regularly to evaluate and classify new wheat varieties for use in both domestic and export markets. Nutritional demands are increasingly important in these discussions.

> And another end of the market is that where wheat is being used for food production, the emphasis on food manufacture and food supply has been driven around health issues, and we see that as being a lasting consumer pull. And so a lot of what we see ourselves being involved with, and a lot of changes for food manufacturers that we believe will flow right back to grain growers, will be how wheat fits into those dietary demands for more nutritional products and how those can be met with new developments. (Kevin)

> I guess at the other end of the scale we're getting a lot more interest in functional foods and breads with particular nutritional attributes. So we're saying to the breeders very much at the moment, 'if you've got wheats that have got particular nutritional attributes that you'd like to show us then we would be very interested to look at them'. (Dennis)

Some of the wheat milled by Hamish travels by train to the Big Starch plant where Jeff works. Flour, arriving at the site at the rate of 1,200 tonnes per day, is converted into what are regarded as primary products – wheat gluten (protein) and starch (carbohydrate). One hundred and forty tonnes of gluten a day is washed and dried, and then sold for use in the baking industry, particularly in the manufacture of bread in the United States, where additional protein is required to 'hold up' the dough. Gluten is also added to vegetarian foods, to pet foods, to aquaculture and other meat industries as an additional source of protein. Modified starches have many traditional food uses including providing texture and 'mouthfeel', enhancing the resistance to freezing and thawing, stabilising foods in various conditions, and providing whiteness and bland flavour.

A range of branded wheat gluten products are made to provide functional and nutritional benefits in foods. These include the ability to increase the protein content of ordinary foods such as soups, beverages and dairy-type foods, as well as sports and health beverages. The products also have the cohesive properties important in foods such as tacos, corn chips and potato chips. Developing these applications requires expertise in, and detailed analysis of, wheat biochemistry, particularly its amino acid composition and protein structure.

Jeff's role is to diversify market opportunities for particular wheat gluten proteins. According to him, the gluten is one of the most lucrative products derived from the flour, selling for as much as A$1,500-$2,000 per tonne.

> A more recent development, which is one that I've been heavily involved in, is developing new proteins from gluten … which we now call 'isolated wheat proteins'. They're a wheat protein which has been modified so that it can behave

as, what we call functional food ingredients, and they're used in a wide variety of applications depending on how we manufacture them. So now, instead of wheat going into the baking industry as gluten, we now have wheat proteins going into the meat industry where they compete with soy bean proteins. (Jeff)

They are added to foods such as processed meat for nutritional supplementation, as a replacement for more expensive meat proteins. They are also used as an emulsifier, to bind the fat in manufactured meat products like frankfurters and meat pies, so that the product can be handled more easily. As Jeff explains, these wheat proteins have particular market and processing advantages in the way that they can be obscured and hidden. They can also compete with Genetically Modified (GM) soy products in an environment where consumers are demanding GM-free food products.

> Our wheat proteins are GM free and the products that we can manufacture will compete with the soy protein type ingredients ... in fact now we're utilising wheat proteins to wholly or partially replace milk proteins in milk, dairy-like foods. We've been able to develop processes which take out all the cereal flavour. Nobody wants ice cream that tastes like the cone, so we've been able to take the flavour away from the wheat protein and produce a protein material which basically is bland and can be used for almost whatever that protein can be used [for] ... (Jeff)

New products like these are continuously in development. Jeff had been quick to see the market potential of glutamine-enriched sports drinks and health foods. Glutamine is an amino acid that occurs naturally in the body, but is now isolated and sold as a nutritional supplement for patients with a variety of diseases, and for sports medicine. It can be sourced in a variety of ways, but as he explains there are advantages in concentrating it from wheat.

> Initially, that glutamine has been produced by chemical synthesis or various forms of biotechnology and we said hey, why would we do that? Thirty per cent of wheat protein is gluten. So we have been focusing on methods which will allow us to present our wheat protein in a form that delivers the glutamine as a therapeutic source, but in a form also that athletes and patients in a medical scenario can consume ... in substantial quantities. (Jeff)

Perhaps reversing the old joke that cornflakes offer no more nutrition than their packet, Big Starch's industrial products promise excellent adhesive properties in contexts from paper to building products such as wall linings. Once again, a combination of consistency and differentiation is necessary to meet the demands of the market and maximise profitability. Thus the wide selection of starches improves the internal strength of diverse papers and cardboards. A specialised series provides maximum ink microcapsule protection for carbonless copy paper.

Assembling this as a product requires a further pulling apart, stripping out starch granules smaller than 10 microns.

'Nobody wants ice cream that tastes like the cone'. The 'naturalness' of ice cream is enhanced by the invisibility of the wheat. In contrast, visible wheat enhances the naturalness of bread. For example, one of Dennis's proudest professional achievements is the creation of a women's health bread that includes soy and linseed as well as wheat. Its structure and ingredient list is just as complex as ice cream, but this bread celebrates its connection to wheat with packaging that denotes an individual farmer hand harvesting a golden field of grain.

Wheat becomes Meat and Milk

Other important products in the supermarket are the various forms of pig meat and milk. Although Australian meat consumption in general has decreased since the 1930s, consumption of pig meat has increased – to 19 kg per capita in 1998–99, a 17.9 per cent increase from the 1993–94 figures (ABS 2000). Australian dairy consumption, on the other hand, has remained relatively constant since the early 1990s, with skim milk now a larger share of the market (ABS 2000). Each of these products can be thought of as a shadow place of Australian wheat, with examples of the ecological transformation also found in other meat and poultry. As the most significant source of energy for both pigs and dairy cows, wheat becomes part of bodies, first animal and then human. In 2005–6 Australia actually consumed slightly more of the wheat it used domestically in the production of stock feed (2.6 million tonne) than was used directly as food (2.4 million tonne) (ABARE 2006). In the USA it takes an estimated 5.9 kg of grain to produce a kilo of pork and 0.5 kg to produce a litre of milk (Pimental and Pimental 2008). In this way coeliacs consume wheat that has been transformed by and into the bodies of animals.

In this part of the wheat network, calories rather than particular qualities of wheat is key.

> We don't care. They're the cheapest wheat we can get. We don't have a portion of particular value as someone like the AWB who have milling wheats and noodle wheats. To us wheat is wheat. Wheat is energy for us. That's what we buy it for. (Allen, Manager, Stock feed mill)

Allen manages the stock feed milling division of an internationally owned food production group, which claims to be Australia's largest integrated pig producer. They mill and produce the feed; breed, grow and slaughter the pigs; and provide meat to wholesale outlets in Melbourne and elsewhere. Piggeries have been historically successful in the Albury-Wodonga district since the mid-1970s, when group farms were established to take advantage of the milling offal, a by-product from flour mills in Albury. In 2006, Allen's mill made about 320,000 tonnes of feed, supplying all requirements for over 1 million pigs produced per year, in addition to the breeding stock owned by the company. The mill also produced stock feed for the wider market, which supplies other piggeries and other business, such as dairy farms.

For Allen, as quoted above, the identity of wheat is simple – wheat is energy. Although there is some substitution of grain types in the various stock feed recipes for cost and availability, the sheer quantities of grain required determine that most of it is wheat. The mill receives grain from trucks direct from farms, then moves it around in elevators and on conveyors; the flows of movement described in Chapter 6 continue on a smaller scale inside the feed mill.

Figure 8.3 Pelleted feed at the stock feed mill

Allen describes the stock feed process as a bit like making a cake, just 'very efficiently'. There is a strong emphasis on maintenance and maintaining smooth operations at the mill. Although the piggeries have some feed storage capacity, this is limited, so the whole feed production process is '24/7'. Pigs need to be fed regardless of mechanical breakdowns or other mishaps. Wheat as both grain and as mill offal is the major energy or carbohydrate component in the feed. Depending on the recipe, wheat and other grains such as barley, triticale and canola, as well as ingredients like tallow and soy, meat, blood and fish meal are pressed and hot mashed in three tonne batches before being pelleted through a press and sieved (Figure 8.3). Recipes are customised for different dietary requirements in the production pig's life cycle. For example, for lactating sows there is 'a high density feed containing the optimum balance of nutrients to maintain breeding pigs in ideal body condition and to maximise reproductive efficiency and piglet weaning weights'. There is a different product 'to maximise lean tissue deposition during the time the animal has the greatest capacity for growth – from 14 to 16 weeks of age or 30–45 kg live weight and until target weight of 60 kg.' A similar range of products is produced for dairy cows. The whole process is automated and computer controlled. The finished product is then collected by truck and delivered direct to the pig farms. All in all, about 6,000 tonnes of feed a week, or 230 truck movements out.

Although the feed is produced throughout the year, harvest is the most important time for the business. Allen reports that they 'go mad', working very hard at harvest time accumulating grain, because this is when it is sourced at its cheapest possible price. This process of accumulation can include physical acquisition, but mostly involves financial transactions to secure shares in the wheat pool. Without accumulation, a constant regular supply of grain to the mill could not take place. This timing makes all the difference in a margin business. They buy the cheapest possible grain, which is the primary determining factor in their profits. If they have to pay more for grain, then eventually, down the line, the price of pork goes up.

> Price of wheat, price of meat. They are two things that drive this business. It's that simple … The two things, the key things that drive people's decision to buy meat is price and whether they feel it's a wholesome product and I think that's the key to it … whether the actual retailer, the person who ultimately buys that animal has any concept that once they were fed wheat, I think that's long been separated. (Allen)

Pigs and wheat are connected in lots of other ways too. For Ted, managing director of Vertical Pigs, the connection between wheat and pigs is based on feeding wheat to the pigs, but it is also about infrastructure, logistics and the design of an agricultural business. The company started out as a trucking and transport business, but over the past 15 years has vertically integrated into the pig and wheat farming business, amongst other things. Today they still operate trucking and transport of grain and feed, but in addition they operate a 175,000 tonne wheat storage facility, they store and supply commercial fertiliser and fuel to local businesses and they produce about 120,000 baconer pigs a year. Like other parts of the business, the piggery is a joint venture project, in this instance in conjunction with the company for whom Allen works. Allen's company owns the pigs, but Ted's is responsible for the feed, the transport and the on-site management.

According to Ted, vertical integration has enabled his company to buffer some of the risks associated with agriculture. For example when the cropping, grain storage, handling and fertiliser parts of the business suffer during drought, the logistics and the piggery provide a more reliable income stream. Similarly, by owning the transport and freight parts of the operation, empty truck loads are negated. The business is not just integrated in an economic sense. The company has also been able to integrate parts of the meat and wheat business in another very physical way. Its pigs are raised in shelters (Figure 8.4) on a dry straw bedding material made of rice hulls. In combination with the straw, the pig manure is composted and made into a fertiliser. This fertiliser is used on the company's own wheat crop, located directly adjacent to the pig sheds. This makes sense financially, as the manure does not have to be transported. But it also connects in this example with the inputs required for the cropping operation, as wheat fertiliser inputs are reduced, and eventually wheat yields improved. This is not meant to downplay or simplify the expertise involved in agricultural production, but rather to illustrate how people have been able to make other connections between wheat and meat, and in the same process meet some of the economic challenges involved in modern agricultural production.

> The pig integration thing – the people of the cities, and not just capital cities, do not, would not believe what's got to happen to get a loaf of bread or a pork chop on the table. They just wouldn't believe it. The amount of people in the food chain and you know, just to make all the processes happen … you know, from all parties concerned, the major significance of it. (Ted)

Figure 8.4 Pig shelters at Vertical Pigs

The terms grain and wheat are often used interchangeably in discussions about pig feed and, although wheat is critically important to the production of pork, it is difficult for a variety of reasons to disentangle the wheat from the other grain. According to Dennis, whose company also produces stock feed, the substitution of grains in animal feed is dependent on the recipe and driven by price and availability.

> Now ... stock feed is very much driven by price and the grades that are available for stock feed can be sorghum, barley, oats, triticale, wheat, probably not much maize because there's not a lot around. So it very much depends on the relative prices of grains versus their nutritional value. All stock feed plants are run by, on what they call least cost formulations. In front of them they have really, all the prices and all the energy values of all the ingredients ... available at any one time and they can very, very quickly dial up and say 'what is the least cost way to make this ration and meet the specification today?' And that can vary from one day to the next ... wheat could be in for one period and then the sorghum has just come off from northern NSW. It comes off about now ... you'll probably find that all the wheat would be displaced and that varies ... (Dennis)

This kind of substitution means that obtaining reliable figures of the amount and proportions of wheat used in animal feed is quite problematic. In 2006, Allen's feed company paid growers the second ever highest prices for wheat (over A$100 per tonne more than in the previous year) because of shortages due to the drought. In fact, 2006 was quite extraordinary because stock feed producers, needing a constant supply of food and energy to the pigs, paid the highest prices available for wheat across the entire market. Thus some wheat qualities, carefully nurtured by farmers from high protein varieties, were 'wasted' in these drought years, albeit at the same time earning their farmer more per tonne than ever, and transferred as energy into pig protein.

The Wheat in Milk

Mick, a dairy farmer on the south coast of NSW, is 'purely and simply after an energy source. We can find protein elsewhere.' Mick's family has been farming on their property since 1839 and his son will be the sixth generation to do so. Mick milks 250 cows on 280 hectares, producing over 2 million litres of milk a year. Things have changed a lot since he began dairying (Figure 8.5). One significant change has been the rapid expansion of suburbia along the fertile, well watered coastal strip that was previously rainforest and then dairying country for most of Mick's family's history. Farmers left the coast as they were paid substantial amounts by developers and moved their businesses inland in search of cheaper feed. Dairying is also a margin business, particularly since deregulation of the Australian dairy industry. Significant mechanisation and genetic introductions from European, North American and New Zealand stock have resulted in improvements in the Australian dairy herd. Instead of 2,500 litres per head a year when Mick first started dairying in the 1960s, Mick's cows now average about 8,000 litres per year.

Figure 8.5 Cows, South Coast NSW

As a result of these changes Mick has had to employ a number of strategies to remain dairying. He has developed continuing relationships with other agents like truck drivers who regularly source and cart cheap hay. He has also developed direct relationships with grain growers, cutting out the 'middlemen' so that he can receive grain without the additional handling and on-costs. He engages a nutrition consultant to advise him about feed requirements in relation to milk output. In one sense, having remained in the industry through such change, Mick has been successful. But he compares what he is doing now to his experiences of this past. In his own words, if he was doing this kind of thing before, 'growing grass like this' when he started, he would have been the 'world's best farmer', but now he's just 'Joe Average'.

During 2006 Mick purchased about 500 tonnes of wheat, which he stores on-site. He owns his own roller mill which he uses to crack the grain to add to purchased pelleted stock feed. He buys commercial wheat derived pellets for dairy cattle, supplemented with biscuit meal (a by-product of biscuit and cereal manufacturing). According to Mick his cows might receive over half their ration in wheat, the rest in grass, hay and silage. Buffers and pro-biotic additives, which stimulate gut bacteria, are required in order for the cows to be fed grain and grain products in these proportions.

It's not hard to work it out ... For example a cow giving 30 litres of milk a day will require approximately, I've got to do the sums in my head, 26 kilograms of dry matter per day and we at times feed up to 12 kilograms of dry matter in the dairy, which is the wheat ration. So they could be getting over 50 per cent of their total ration from purchased grain, purchased feed, whether it be in the pellets or in the straight wheat. (Mick, dairy farmer)

The rations do vary, according to season, availability and price. In the winter, rations need to be supplemented, because the paddock grass growth slows down and the cows need additional feed as they burn energy keeping warm. (Australian dairy herds are generally not housed indoors during winter.) Rations also vary because of the way in which Mick is paid for his milk, which is according to the protein and fat content. And so depending on the cow, he uses the nutritionist's services to adjust the diet to the extent possible, to produce the most profitable proportions of protein and fat content in the milk, for a given volume. 'Barley, triticale, oats. Wheat is the grain with the highest energy value. And in fact given we may get back one day to normal seasons we would probably split the grain 50/50 wheat, 50/50 triticale. And triticale is actually a derivative from wheat anyway'(Mick, dairy farmer).

Wheat energises the system, but rather than being unidirectional, it moves around, re-energising other parts of the network, fuelling and feeding as it goes. As encapsulated energy, it can be stored and pooled, moderating the effect of the seasonality and variability of drought. Its mobility and energetic power allows Mick to stay on his land and farm in the face of considerable economic pressure to do otherwise.

Conclusions

Mick's cows and his verdant pastures are visible as you drive south from Wollongong along the new freeway. They contribute to the rural amenity of the region, even if most people speeding south on a Friday evening are more focused on the beaches and forests of the coastal strip. If as consumers we know that cows eat grass to make milk, most of us are unaware that it is the grain, and the wheat, that enables the milk to be supplied consistently and conveniently, whenever we might need it. We have used three intertwining themes in this chapter to follow some wheat transformations: consistency, craftability and in/visibility. If an important conclusion of Chapter 6 was the complexity of the structures and processes that are necessary to make wheat journeys as frictionless as possible, a parallel here is the deconstruction (both physical and chemical), reassembly and crafting necessary to deliver consistent food and consumer goods to mostly urban populations. Wheat is uniquely placed to contribute to this edifice through its particular and starchy materiality.

Wheat has a much more fixed identity as a human food than as non-food products. This identity is fixed in markers that denote both presence and absence, facilitated by regulatory regimes concerned with the health of the human body. As the healthy staple food of the nation, wheat is marked and celebrated via imagery of grains, stalks, sheaves and golden fields on food packaging. It is presented as a product of nature. Consumers understand themselves to be eating wheat, an unmediated product, often presented with a visual connection to the farmer and the paddock (but never the truck or the factory). Even the highly crafted soy and linseed bread masks its crafting with images on the package that connect it back to nature. The human body is understood as the right place for this food, unless of course the human is a coeliac sufferer. In that case wheat is dangerous and must be kept outside the body. But its identity is just as fixed, in fact more so because of its dangerous qualities. These qualities are what have rendered it visible in new labelling regimes designed to protect consumer health. In the experience of coeliacs wheat is 'surprisingly' everywhere, but also, not really food. Instead it must be located and avoided.

Outside the regime of human food, wheat is a much less stable category. In non-food products within the supermarket it is not even labelled as existing, let alone as wheat. For manufacturers, food and industrial scientists, the malleability and unstable identity of wheat is what they value about it. It is easy to hide. Its capacity to be broken down as different constituent parts, and recrafted into other things is fundamental. In industrial processes in laboratories and factories, the constituent components of wheat grain are broken apart, regrouped and reconstituted into new collectives; stock feed, functional foods and isolated proteins. Thus the pet food manufacturer celebrates the gluten which allows the kibble to be extruded into shapes with brand identity; the stock feed miller and the dairy farmer value wheat as energy; the starch scientist separates wheat proteins that enhance the mouth feel and texture of meat pies and ice cream. These food crafters further challenge the analytical distinction between consumption and production, being both consumers of the farmer's product, and producers of new goods for consumers in supermarkets and elsewhere.

There are several illusions at play here. The ongoing elision of wheat for food as a fixed category, corresponding to 'nature', fuels the expectation that food will and should be cheap. As a product of nature people expect to pay very little for it, and will likely resist internalisation of environmental costs, for example via carbon trading. Nor are they likely to be able to discern, or want to pay the farmer for, a higher quality of generic wheat. The humble food label, on the other hand, would never be taken for a natural entity, rather it is more likely to be seen as emblematic of industrial agri-food businesses, the list of three-digit food additives the butt of many jokes. But it can be considered one of the 'matrix of knowledge practices that must be re-aligned if the overstretched fabric of trust transacting the distant intimacies of growing and eating is to be rewoven' (Whatmore 2002: 144). The one we have focused on, labelling for wheat presence, grew out of a human health issue, but others have emerged from the demand for different types of accountability – origins, organics and transparency of pricing. The food label is a quiet hero in helping us to understand the shadow places of wheat.

Conversely, the paddocks of golden grain are as much industrial landscapes as they are food production landscapes, providing fibre and starch that packages and structures our lives in many different ways. We present this different way of seeing wheat, not to downplay the importance of understanding broader rural landscapes, but rather because they are so important. Discussions about the sustainability of wheat production in the face of climate change, in Australia as elsewhere, will need to consider the diversity and complexity of wheat journeys and transformations. What futures for those landscapes are possible, desirable or sustainable in the face of climate change and other challenges? It is to this question that we turn in the following chapter.

Chapter 9
Wheat Futures

Next Year's Country

> This is called 'next year's' country. It's always, 'What about next year?' (Keith, household F)

In telling us how a large grower in the district was in huge financial difficulty from having forward sold a crop that did not eventuate, and yet was ready to 'go again' next year, Fiona's husband Keith described the farming landscape as 'next year's country'. Over tea and the gluten-free biscuits, numerous examples emerged. Each combined stubborn optimism with a temporal horizon that focused on getting a crop in the ground 'next year', juggling debt levels and input costs in anticipation of the elusive bumper harvest, or even the 'good enough' one that would enable them to survive until the year after that. There was the 'young bloke' down the road who had sold his farm to his brother so the banks 'wouldn't crunch him to death'. Keith and Fiona themselves had pre-bought a quarter of their fertiliser at the unprecedented price of A\$850 per tonne, thinking the price would go even higher.

> What we've got this time round will probably only give us enough to go back to the same debt level as what we've got now, which is the highest it's ever been since we've been farming. So what do you do? … You go back again don't you? There's got to be greater margins, there's got to be a way. (Keith)

> Farmers are always good at going back. (Fiona, household F)

Several hundred kilometres away, the same bleak, rather dogged, focus on next year was articulated by Don;

> one thing I've found out for sure and for certain as long as I've been farming [is] that, okay, last year was a bad year. This year, we don't know what it will be like but it won't be like last year, it will be different … it may be worse, it may not be. It may be better in places and worse in other places. You just go along, when the breaks come, yeah. (Don, household A)

Agriculture has been an inherently risky enterprise since hunter-gatherers traded diversification and mobility for monocultures and sedentism. Reminding us of the stubborn physicality of the plants themselves, Moschini and Hennessy (2001: 89) argued that agricultural risk is 'heightened by the fact that time itself plays a particularly important role in agricultural production, because long production lags are dictated by the biological processes that underlie the production of crops and the growth of animals'. But that seasonal cycle is only one of the temporalities in play. How are farmers – or any of us – to think about the longer term future of wheat farming, against an impossibly complex set of global processes, when they need to make Next Year work to have any chance at a viable longer term?

One of the financial means that NSW farmers use to hedge against uncertainties in their own crop is to buy wheat futures. Like other 'futures' on international exchanges, these are standardised, centrally cleared and deliverable contracts for specific quantities of a commodity at a specified price, with delivery set at a specified time in the future. They have full or partial fungibility. For example, the commodity Australian Milling Wheat as offered on the Australian Securities Exchange under the code 'AWM' is deliverable as a grade of APW2 or better, with a protein level of at least 10 per cent. Higher levels of protein attract a premium of 60 cents per 0.1 per cent, up to 11.4 per cent (ASX 2011). As we saw in Chapter 6, these standards and measurements of quality mediate the physical flow of wheat. In the context of futures they again facilitate fungibility, the seamless flow of information, money and physical wheat in and through something called the global economy. The apparent seamlessness is in stark contrast to the spatial and temporal variability of the crop on the ground, and the risks that determine how much of it will be actually harvested from the ground, as the phone call between Fred in the paddock and Bloombergs in Melbourne showed in Chapter 4. This of course is precisely what makes them attractive investments to farmers who want to hedge against the uncertainties in their own crop, and to speculators who are trying to make money out of risk.

In this chapter we consider what a focus on the farming household, and its interaction with the material plantiness of wheat, can offer to our understanding of how the complex challenges of the next few decades might be negotiated. The tensions are many; between Next Year and several decades hence, between increasing demand for wheat and increasing difficulty maintaining viable farming households, between protein levels in the endosperm of a wheat grain and computer manipulations of the same in Chicago or Singapore.

And the challenges are many. It is argued that the world will need to produce 70 per cent more food by 2050 to meet rising demand (FAO 2009, Carberry et al. 2010). The impact of anthropogenic climate change on global agriculture, while regionally variable, is likely to be negative overall (Nelson et al. 2009). In a context of population growth, climate change overlays a new set of challenges on longstanding risks to food production (Alcamo et al. 2007, Ortiz et al. 2008), and interacts with issues such as burgeoning demand in the developing world, food security and biofuels (Burton and Lim 2005, Howden et al. 2007, Tubiello and Fisher 2007). Peak oil or peak phosphorus (Cordell et al. 2009) may yet force threshold changes before climate change does. Since the global financial crisis of 2008, financial institutions have increasingly turned to commodities as real estate and other investments became less viable, further contributing to the financialisation of agriculture and complication of food security (Ghosh 2010). The sharp rise in food commodity prices in 2007–08, triggering riots in a number of parts of the world, reflected the interaction of many of these factors, particularly a long term decline to low stock-to-utilisation ratios, and EU and US policies to increase the use of biofuels (Keyzer et al. 2008, Piesse and Thirtle 2009). Most scholarly overviews have argued that major productivity increases through scientific breeding advances (including probably genetic modification), in association with more mechanisation and improved economies of scale, will be necessary to meet these challenges in the decades ahead (Dixon et al. 2009, Carberry et al. 2010, PMSEIC 2010). Although a third of world grain production (and almost 70 per cent of grain production in developed economies) goes to feed animals, changing the dietary patterns of the affluent or the becoming-affluent is not widely considered to be a possible area of policy traction.

Our entry point to these tightly knotted tensions and challenges is climate change and wheat farming households. We are specifically not arguing that farmers will or should bear the brunt of climate change response. Nor do we argue that climate change mitigation and adaptation should be understood only in terms of responding to climatic variables such as rainfall and temperature; indeed we explicitly argue for a conceptualisation of an assemblage that comprises 'more-than-climate'. Rather we use this particular entry point to analyse the ways localised experience and global processes are furled together. Understanding some aspects of these relations opens up new ways to think about intervention for change.

Following the call of Mike Hulme (2008) to examine cultures of climate, this study joins the growing body of literature viewing through a critical lens the question of climate change adaptation. Most adaptation studies in agriculture, particularly in the developed world, have focused on agronomic and top down perspectives. We contend that these perspectives must be complemented by a more fine-grained perspective that pays attention to everyday life. Adaptation to climate change in Australian wheat farming, however it occurs, will be undertaken by the almost 30,000 farmers who grow it (ABS 2006). Big decisions about food production and landscape change across significant areas of Australia are being made by individuals and households. For a farmer in the wheat belt of NSW, global climate processes take expression in local processes such as the timing and intensity of the autumn break, the reliability of winter and spring rains or the presence/absence of frost. But further, 'climate change' has expression in non-climatic ways, for example via public discourse, media hype, scientific controversy, wheat futures trading, carbon sequestration discussions and so on.

Australian Wheat Farming Households and Cultures of Climate Change

Most models suggest that southern Australia will become drier and northern Australia wetter by 2030, a trend that has already become visible over the last few decades. Projections in total productivity for Australian wheat at 2030 are regionally variable, and include predictions of both decreased and increased productivity under different rainfall scenarios (Howden and Jones 2001, Heyhoe et al. 2007, Wang et al. 2009). While the risk of significant decreases in yield is high in regions such as Western Australia, beneficial impacts on yield are predicted for parts of NSW (Howden and Jones 2001, Fairweather and Cowie 2007). There is considerable uncertainty about the role of enhanced CO_2 and its interaction with rainfall and temperature changes (Ludwig and Asseng 2006). Associated detrimental impacts are also predicted, including decreases in grain protein content, increases in pests and diseases (Basher et al. 1998), and changes in salinity and erosion (Howden and Jones 2001). As Harle et al. point out, climate change will interact with climate variability in two main ways.

> First, many of the impacts of climate change are likely to be through changes in the extremes of natural variation (higher peak temperatures and fewer frosts) rather than as a result of changes in average temperatures. Second, climate change models predict that climate variation will increase with climate change … (Harle et al. 2007: 75)

It is of particular importance for Australian wheat that the rainfall decreases are projected to be greatest in the growth seasons of winter and spring (Gunasekera et al. 2007: Table 2). Howden et al. argue that the challenges of adaptation to changes already in train are urgent. Drawing attention to the importance of scale in decision-making processes, they argue that

> Short-term climate adaptation by farmers may be accomplished by taking into account local climate trends if there is a strong correspondence between these trends and projected climate changes, or it may be via climate forecasting at scales from daily to interannual. However, farmers may find limited utility in long-term projections of climate, given the high uncertainties at the finer spatial and temporal scales at which their decisions are made. (Howden et al. 2007: 19692)

As Howden et al. acknowledge, and as we elaborated in earlier chapters, most detailed decisions about purchasing, variety selection, planting and so on are made for the year ahead only, and within constrained timing windows. It is widely recognised that many climate adaptation decisions at the farm level are variations on existing risk management processes, including climate risk (Howden et al. 2007). The decisions under discussion are mostly agronomic and economic, such as altering varieties that are planted, altering timing or location of cropping, and improving the effectiveness of pest and weed management, among others (Heyhoe et al. 2007, Howden et al. 2007, Lobell et al. 2008). Australian farmers are generally considered to have high adaptive capacity because they have long had to deal with a highly variable climate (Heyhoe et al. 2007).

Adaptation measures in the form of different varieties or changed planting windows have been modelled 'to reduce the impacts of climate change by almost 50 per cent' (Heyhoe et al. 2007: 175). Our point here is not to dwell on the actual percentages. Rather we are interested in exposing and exploring the sociocultural complexity hidden in the blank space between two lines on a graph, labelled 'unadapted' and 'adapted'. Sociocultural dimensions of adaptation have received most attention in relation to the developing world, where communities and nations are recognised to be particularly vulnerable (Adger et al. 2003, Ziervogel et al. 2006), and also in relation to indigenous people (Ford et al. 2008). Relatively wealthy well-educated countries are often assumed to have strong adaptive capacity (Brooks et al. 2005), leading to a focus on technological dimensions of adaptation, such as agronomic change in the case of agriculture. On the other hand, many parts of the developing world have great resilience and adaptive capacity (Coulthard 2008), and well-established institutions may lack the flexibility to respond quickly.

Diversity in vulnerability and resilience is increasingly recognised within broader social categories as well as between them (for example Acosta-Michlik et al. 2008). Our study responds to calls for more attention to be paid to the social context of adaptation in the developed world (O'Brien et al. 2006, Gorman-Murray 2010). This means drawing on research methods and approaches that have been more commonly used in the developing world (for example participatory approaches, Kelkar et al. 2008). It also means more attention to the household scale of analysis (Thornton et al. 2008), and to how climate change interacts with other drivers, including socioeconomic ones (Wei et al. 2009).

Risk and Climate Change

We did not set out to interview the farmers about climate change in particular, but about their daily lives and their embodied interactions with wheat. However there was a significant shift between the December 2006 and December 2007 fieldwork – climate change had become more prominent as a topic of national conversation in Australia, reflecting international landmark events such as the Stern Review and Al Gore's *An Inconvenient Truth*. Australia's failure, with the USA, to sign the Kyoto Protocol under the conservative Howard government was a hot topic in the Federal election campaign in November 2007. Indeed the first official act of the new Rudd Labor government was to commence the process of ratifying the protocol. In December 2007 we made a point of explicitly asking about climate change in our follow up interviews.

Even when a farmer professes 'belief' pro or con climate change, there is still considerable uncertainty over whether the current drought is a manifestation of that, or just another drought. The following responses exemplify the range of views expressed about climate change:

it's happening but I'm not blaming the drought that we've just had on climate change. (Chris, household B)

Well I sort of wondered for a while but I think I've got to go with it a bit now, I think something's going on isn't there? But then you talk to older people, some of the older fellows and ... the 30s were like this. (Jim, household M)

I'm not too big on climate change ... A lot of people use it as an excuse for failure. (Ted household E, and manager of Vertical Pigs)

I think it's climate variability and it's always been and it always will be ... mother nature, she overrules everything. (Joseph, household G)

Intellectual understandings based on reading or media engagement are interacting here with embodied experiences drawing on memory of self and others. As to what can be done about it, the more entrepreneurial farmers tend to see opportunities rather than threats, for example 'there'll be lots of opportunities but you just don't know with the climate what's really going to happen. It's definitely warmer' (Andy, household O). Further, decision-making is mostly still framed through climate variability, and the necessity of living with it, rather than something called climate change.

The extremes of droughts and storms are going to be greater but if you're going to be a farmer you've got to be able to handle that variability. (Chris)

And we'll do what we've always done, we'll adapt because we have to ... that's what farmers do now, we work out the weather every day and try and do the best with what mother nature deals a hand. That's our job description. (Joseph, household G)

The price of seed and fertiliser interacts with underlying soil moisture, the timing of the autumn rains and international grain prices, among other things. Decisions are framed through the temporality of 'Next Year'; as the following discussion indicates, sowing decisions need to be made each season on the basis of the current information.

Jim (Household M): Yeah it's [climate change] definitely probably going to affect our management decisions in the next few years ... Even if we do have it dry all year you still have to think about management decisions and what you do at sowing time, it's not something you can do gradually as you go along. Those decisions have to be made right at the start ...

Interviewer: So it doesn't necessarily make you more cautious overall?

Jim: If you had a list of management decisions you had to be thoughtful of next year it wouldn't be the top one … Probably nutrient management is number one. It all comes back to trying to guess the climate, I guess, but knowing how low we can go with our inputs without penalising the outcome at harvest next year.

Similarly, Terence (Household J) discusses how he will adapt if there is increased summer rain:

Being a predominantly winter rainfall area here, part of our problem over the last couple of years has been that we've had no summer rainfall. Now if we get summer rainfall and build up the moisture reserves in the soil, we could have tolerated the dry period we had this year in August, September, October a lot better if we had've had more subsoil moisture. So what I'm saying is that's a way we can manage it if we have that subsoil moisture from the summer rains, control our summer weeds and maintain our straw coverage and just keep it there.

Interviewer: Your soils will hang onto that moisture?

Terence: Not as well as some soils, but yes they all do a bit. But obviously the heavier clay soils will hold it better than what our soils will but we'll work at it and I guess if we get climate change, our environment will probably change a little bit with it too and we'll work with it. We've got no choice.

Don (household A) talked about the interaction of evaporative regimes with seasonal changes:

We get more evaporation so the rain that does fall disappears into the atmosphere quicker so that's going to limit our yields. Whether [in] fact you're actually having less rain or more rain, I mean the fact that it's hotter means that yield potential is capped a bit more, but if we end up having drier winters and springs which is when we're trying to grow our winter crops it's going to make a big difference to what we can grow.

The issue of connecting adaptive capacity to the temporalities of farm-scale decision-making was also addressed by state government agronomist Kate, who is responsible for providing advice to farmers across an area of about 6,000 km^2 that includes Keith and Fiona's farm. The recent climatic variability has meant that her work is increasingly focused on assessing crop failures and reporting on the moisture stress scenarios for crops across the district. Her own agronomic experience over the past few years has been in a series of really abnormal situations, and she is not quite sure any more what a normal year looks like. As well as drought and disease, climate change has become part of her discussions with growers.

Kate was most concerned to point out that while she had relative security in her publicly funded position, recent years had seen very real struggles for many people in the district, which she described as being a fairly traditional and conservative growing area. She saw part of her role as alleviation of community stress and anxiety by having good technical and agronomic knowledge and communicating that well to growers. As well as writing up the variety and agronomic trials, she is constantly reading and summarising new research publications for grower updates and seminars. Her role is also extremely practical and hands on, including farm inspections and laboratory testing of disease outbreaks.

> Yeah, it's really hard because everyone knows about it [climate change] and they're all panicking about it but there's nothing we know [t]hat we can do for a solution ... Half the time it's just thinking through, ok well if we have later and later breaks then it's going to need that later and later varieties, and you're going to have to switch to systems that maximise water retention and all that sort of thing, which might be changing your machinery. Half the time it's just thinking it through. It is a slow process climate change, I guess that's one good thing, it's not like stripe rust where you say right we'll have to make a change next year, this is all the options we've got ... So it's being fairly adaptable and being able to think through scenarios and think on your feet. Talk to a lot of people. (Kate)

The connection to stripe rust in this discussion highlights that farmers have to regularly make major variety switches in order to deal with rust problems. In this sense they have strong existing resilience, but it takes expression through very specific temporal windows.

Durum growers in northern NSW were specifically asked whether they had made any changes to their business or on their farm as a result of climate change. Most were aware of the issue but concern about it and the desire to do anything themselves varied. One grower said the whole thing was a blur in the information overload. Four out of the five said they had done nothing specifically but one reported that he had changed to skip row planting[1] in anticipation that his farm would experience drier conditions because of climate change, even though he had not seen anything yet that worried him on his own farm. Two growers said that the issue does worry them because it implied a changing environment, so their general strategy included improving overall farm practices, which to them meant conserving soil moisture and minimising soil damage. As we saw in Chapter 5, durum growers in the far north may have both more options (two crops a season, more soil moisture and fertility) and more risks (high costs of soil preparation and management, risk of summer rain at harvest) than farmers further south. Further, recent research providing more detailed regional projections (Wang et al. 2011)

---

1   A sowing configuration in which only every second row is sown, with the aim of conserving soil moisture.

does not necessarily suggest the rainfall prognosis is better for durum areas, however areas with better soil moisture profiles will be better placed.

Climate change is a contentious issue in the region because of the substantial coal deposits and associated mining operations in the Liverpool Plains. Government infrastructure projects have upgraded both rail and road facilities across the area to support coal development. One grower's response to the question about climate change, which they felt was reflective of a wider sentiment amongst the plains community, was that government was hypocritical in attempting to constrain carbon emissions while also benefiting financially from investment in the coal industry.

These examples illustrate several connected things. First, they reiterate the detailed ecological knowledge that is part of any farmer's life. It is second nature to them to observe the interactions of soil, soil moisture, rainfall, evaporation and seasonal changes, among many other things. Second, whatever they actually think is happening about climate change (that is, whether they 'believe' in it or not) is only partly relevant to the processes by which they mediate seasonal climate variability into their daily lives. Thus, any policy strategies that aim to simply educate farmers about the 'facts' of climate change will likely miss the point. Third, Australian wheat farmers already have many necessary skills and capacities to deal with risk and uncertainty, capacities that vary primarily due to education, family structure, socioeconomic status and geographic location with respect to rainfall and transport options. Fourth, the same strategic and reactive approaches to risk that we discussed in Chapter 5 underpin farmer approaches to climate change risk. Although there are not neat correlations between different types of risk, our data shows that different types of risk interact, and tend to coalesce into total packages of vulnerability (reactive approaches) and resilience (strategic approaches). Vulnerability to climate-related risks parallels that to social and financial risks. For example, it is only strategic thinkers who are even contemplating the possibility that climate change will provide positive opportunities such as vegetation offsets or stewardship payments (cf. Hatfield-Dodds et al. 2007). And when the interviewer suggested to a very strategic family that they sounded well prepared for the future, they replied, 'As long as it rains'.

It is clear from this evidence that adaptation research needs to include more fine-grained sociocultural approaches that consider the implications for the farmer, as well as the crop. Following Howden et al. (2007), this needs to be considered separately from, but as complementary to, longer term investments by governments, for example in plant and animal breeding programs, quarantine and research systems, and other infrastructure. It is also important to emphasise that this work captured climate change discussions among farmers at an emergent time in the national discourse. This is valuable in that farmer thinking was clearly in a process of change, but it would be inaccurate to summarise these conversations into firm categories of attitudes or practices. It will be important for future research to document shifts and consolidations over time.

More-than-Climate, More-than-Farming

Of course climate change is not only an issue for farms and farmers. Nor is it only about drought but encompasses increased frequency of extreme events such as storms, floods and cyclones. The morning we interviewed Duncan the manager of the Northern mill, for example, there had been a major power outage due to a large storm the previous night. As well as resetting power to the mill, staff were in the process of cleaning up stormwater debris strewn across the loading and storage floor. Flooding on roads and rail lines was also affecting transport of grain into and out of the mill, disrupting schedules. During the year we were writing this book, areas that had been in drought for a decade experienced floods and plagues (both locust and mouse) of biblical proportions.

Yet, in a similar way to the conceptualisation of hybrid assemblages as being 'more-than-human' (Whatmore 2002), farming households of the NSW wheat belt challenge us more broadly to consider climate change as 'more-than-climate'. We have examined a climatically vulnerable agricultural process, winter wheat cropping, during a period of drought unprecedented in living memory. Even in such a context, 'climate change' is not expressed or experienced 'separately' to anything else. Climate change will have expression in localised and temporally specific weather processes recognisable in the present. And further, it will also have expression in assemblages comprising 'more-than-climate'. These include carbon trading schemes, altered financial instruments, fluctuating prices of inputs such as fuel and fertiliser, public discourse and legislation and breeding programs.

The more-than-climate assemblage operates at a range of interacting scales. At the farm household scale, our research shows that climate risks are filtered and managed with a range of other risks in a total network. There are strong parallels here with the way urban households negotiate climate change and sustainability issues in the context of their everyday lives (Gibson et al. 2011, Waitt et al. 2012). While globalised in sensibility and economic connection, the farmers' management of risk and uncertainty is embedded in the social intricacies of localised daily lives. They recognise a number of different types of risks, and express different ways that climate risks are inextricably intertwined with others, especially financial. This includes not only engagement with climatic phenomena (rainfall, soil moisture, frost and drought) at daily and seasonal timescales, but also day to day practices of information-gathering and filtering.

It is intellectually important for research to resist monolithic constructions of climate change and adaptation before they become too entrenched. Just as important, attention to the fine grain of everyday life will help us make more nuanced and realistic contributions to meeting the multiple challenges of wheat futures.

Chapter 10
Conclusion

Culture and Nature in Cabonne Country

The rural shire of Cabonne, covering a number of towns including Milltown, welcomes visitors to 'Australia's Food Basket' with a sign depicting agricultural abundance (Figure 10.1). Stylised ears of wheat sprout from the top of the basket, at the pinnacle of a landscape providing a harvest of meat, wool, fruits and vegetables. The sign is framed against white-barked eucalypts – icons of Australian nature – in the forest along the roadsides. This image captures two of the paradoxes and contradictions in contemporary thinking, about wheat specifically, and agriculture generally. The paradoxes and contradictions go to the heart of the way agriculture is implicated in – indeed was formative of – the Western culture/nature binary, as Knobloch (1996) has argued.

Figure 10.1 Cabonne Country: 'Australia's Food Basket'

Throughout the book we have seen a number of instances of the way wheat is conflated with nature in being presented as obviously wheaty food – bread, cereals, biscuits and so on. This ignores or hides the profound amount of cultural activity, and the many technological others, required to get that 'nature' to the supermarket or table. It hides the costs in food production and transport – the costs of friction, translation and standardisation. Those costs can be just as complex, at least under the configuration of the current dominant systems, in Small Bread or Small Pasta as in their Big equivalents. As Ted the vertically integrated pig producer said, 'the people of the cities - and not just capital cities - do not, would not believe what's got to happen to get a loaf of bread or a pork chop on the table. They just wouldn't believe it.'

On the other hand wheat is thought of as a cultural product when it suits – when we want to assume all human credit for agriculture, or turn wheat into a commodity with no material difference to coal, oil or gold; or in the Australian context, when we understand agriculture as something of a state of war with a hostile nature, whether drought or flood. Main (2005: 123–5) offers a number of recent examples of this thinking, from both politicians and ecologists. Such views ignore or hide the multiple nonhumans which underpin human existence on the earth. Or to put it another way, they ignore the dependence of agriculture on flourishing natural systems and processes, from groundwater to bees. 'Agriculture is an ecological enterprise that depends on ecosystem processes and functions – such as soil formation, nutrient cycling, maintenance of hydrological cycles, pollination of crops – which are driven by interactions between elements of biodiversity' (Williams 2001, quoted in Main 2005: 125).

The concept of plantiness has helped us think through and make visible some of these underpinnings, not only in agricultural systems but also in contexts of food and industrial production and consumption. We have considered the many collectives and identities under which wheat gathers, resisting easy categorisation. The kind of agency wheat exerts while growing in the paddock has both similarities and differences to what it does in the starch plant. We have extended Matthew Hall's arguments beyond the realm of autonomous individual plants living in spaces free of humans. This dramatically extends not only the plants requiring our ethical engagement, but the spaces and places in which they are found. Wheat landscapes are fibre and industrial landscapes as much as they are food landscapes, and could become more so in the future.

Figure 10.1 reminds us that the boundaries of belonging are spatial as well as conceptual. The welcome to Cabonne Country both crosses and reinforces boundaries. We are being welcomed into a country where both eucalypts and agriculture co-exist, but in a way that suggests they occupy separate spaces; there are no eucalypts depicted on the rolling hills of the sign.

Webs of Wheaty Relations

Both the above paradoxes remind us that Western understandings of nature have usually excluded humans. Yet we are all in this together, we are all implicated. The tight knots of material connection that we have traced, and the complexity of the resulting assemblage, defy the boundaries between native and non-native, vegetation and crop, or the supposed cultural distance between urban and rural. In one sense this is a truism for a conclusion; in another it is a profound challenge to think in terms of associations rather than separations, between different groups of people, and between people and other-than-humans.

What might it mean, for example, to consider wheat as a plant that now belongs as part of Australian nature, to welcome it into Australian ecosystems (as distinct from its current accommodation as belonging to the nation, or the economy)? To accord the wheaty assemblage a kind of ecological belonging would institute a broader sense of its connectedness and underpinnings. It would bring with it a stronger responsibility to establish and maintain appropriate relations of care, and to acknowledge the multiple ways agriculture was embedded in the colonial project. Main has used the concept of regenerative agriculture (rather than sustainable agriculture) to 'acknowledge a painful history of suppression, fragmentation and disorder' (2005: 245) in the Australian context.

It would also mean that we have to consider those humans who are part of the wheaty network – which is all of us – as also belonging. As we and others have discussed, questions of belonging and indigeneity in Australasia have been particularly challenging because inflected through the comparatively recent temporal horizon of European colonisation (Barker 2008, Ginn 2008, Trigger 2008), by comparison, for example, to the Swedish context where cows can be considered by dint of long occupation to 'belong' in nature (Saltzman et al. 2011). There are inklings of such belonging in diverse Aboriginal responses to introduced animals and plants, which demonstrate 'an active intellectual incorporation of some species into Aboriginal cultural traditions' (Trigger 2008: 640).

To be welcomed, or to allow ourselves to belong, in such a way brings with it responsibilities. In the first instance human humility towards its absolute dependence on the planty products of photosynthesis is in order. Further, we need a sense of shared responsibility for decision-making about sustainable food production and landscape management, for our shadow places. Vilification of farmers as environmental vandals is not the way to go. Urban Australia needs a better settlement of sorts with its food producers, and a recognition that the industrial and retail processes are also embedded within these ecologies.

Perhaps the most important responsibility is that of living with, and making decisions within, complexity. Superficially attractive ideas such as local food are rarely the most sustainable solution for highly urbanised populations. The concept of natural food offers false consciousness; it glosses over too much complexity that needs to be pulled apart and understood. By the same token the rhetoric of relentless productivity growth is a discourse from another age.

Relational approaches help us go further, we contend, by showing that the mechanisms of connection and therefore the suggested points of intervention may be different from what we have previously thought. Thinking about assemblages and networks helps us direct necessary attention to strengths and weaknesses in the mechanisms of connection, the points of friction, the thresholds of change. It is different from being romantic or simplistic – we are talking about recasting modernity rather than revisiting arcadia. Fragile connections, for example, might include the administrative and business burdens on farmers, and the 'just in time' production process at Big Bread in Sydney. Policymakers could pay more attention to the demands implicit in the huge piles of paper (and their electronic equivalent) being built up in home offices on farms. (Just in time production has been identified as one of the threats to food security because of the speed at which disease or contamination can be spread through the system (PMSEIC 2010).)

On the farm, an important temporal filter is the seasonal cycle, particularly the planting window. Whatever decisions the farmer would like to make about the long term, say a decade or two hence, have to be mediated through surviving this year and next year. Financial considerations, mainly the structure of debt, are as important in this process as knowledge of appropriate wheat varieties for the predicted season, or effective farm management practices. What sorts of financial structures and processes might we be able to put in place to enable farmers to ride through the worst seasons without having to damage the land further, against their own and community interests? One answer to this is that agriculture is necessarily moving towards a regime of much larger corporate farms, which can better ride out this variability and vulnerability. The loss of sociality, and reductions of human presence, care and knowledge in rural areas entailed in that scenario seem to us to be an undesirable outcome. We are not persuaded that the best way for humans to live on the Australian continent is to manage it from urban coastal fortresses. There must be better ways to organise our subsistence.

Strengths we have identified include farmers' knowledge of their land. The strong existing capabilities in adapting to drought conditions, in being flexible and resilient, offer a resource to the wider community in thinking about the challenges of climate change. Pride in 'quality' throughout the system, including consumer desires for good food, provides a ground for collaborative conversation rather than conflict. And of course wheat itself has great strength – as a flexible and variable plant that fuels and energises many other components of the assemblage. Our relational approach has also pointed to the role and agency of other-than-humans in the human-plant connection. It is worth pausing to consider two of these; the particular mediations of drought and of quality standards.

The Productive Agency of Drought

Between February and May 2010, the Southern Oscillation Index flipped from the dry phase to the wet. Heavy rains became widespread that winter, and TV journalists took choppers over the inland waters to communicate the renewal of the inland to urban audiences. In November national newspaper *The Australian* published a special feature on the breaking of the apocalyptically named 'Millennium Drought'. Many features of that coverage echo the sort of adversarial account critiqued by George Main above – that Australians are locked in battle with a fickle and hostile nature.

Yet, there have also been many inklings of a shift in underlying consciousness and practice. The media documented examples of changed tilling practices, reported that farming households had been quick to seek off-farm work before they got too deep into financial difficulty, and argued that in the end the drought had had the positive outcome of weeding out the most vulnerable farmers (*The Australian* 13–14 November, 2010). Urban TV audiences had had their own drought to contend with, as their regional water catchments dropped to frighteningly low levels, reported weekly on TV weather forecasts along with the Southern Oscillation Index. They cut water consumption in homes and gardens under the influence of water restrictions, installed water tanks and debated recycling and desalination. In Melbourne people grieved the loss of majestic European street and park trees that had survived for a hundred years. The substantial investment of time that farmers put into environmental protection works during the drought (summarised in Chapter 4) foreshadows what agricultural life in a climate change world might look like. Most Australian wheat is rain fed. When it rains, we get a crop. When it does not, we are not using water inefficiently (except in the few areas where wheat is being irrigated, which seems to make little sense on any environmental calculation). Just as the boom and bust cycle of arid ecosystems was celebrated by the TV programs showing millions of birds breeding in the inland wetlands, can we not think of wheat as part of this cycling?

Both urban and rural examples show us a way to value drought as a pause in the life of the nation – to see its agency as productive as well as destructive. Just as drought was productive in the evolution of grassy plantiness millions of years ago, and in the move of wheat from coastal New South Wales to inland areas in the 1860s (Chapter 3), so it has a role in helping us make contemporary and future transitions. Market research showed that in mid-2009, while Australians thought 'global warming' was the most important environmental issue facing the world, they thought 'water management' and 'drought' were the most important issues for Australia (www.roymorgan.com: finding number 4388). That drought provides the lens through which both urban and rural Australians understand wider climatic processes has both positive and negative potential. By mid-2011, climate change had taken over from drought as the main environmental issue facing Australia (www.roymorgan.com: finding number 4677). It remains to be seen whether the wet phase dominant at the time of writing will engender temporary but disciplined celebration of water abundance, or complacency that wet conditions are 'normal', and everything will be alright after all.

Quality and Standards

The measurement of wheat quality starts when a farmer stands in the paddock and grabs a handful of the crop, and continues when the first truckload goes to the receival point. This process starts by focusing on the materiality of the plant, or more specifically the levels of protein and moisture in the grain. But along the way it draws in a much wider constellation; the paper trail, the system of accountability, the notion of standards in food labelling, the certification of organics. These are all crucial and undervalued intermediaries in the system. Because we cannot all grow our own food, we have to be able to trust our systems. This means accepting bureaucracy and regulation and cost. In future we are going to need different ways to measure things, and need to measure different things (footprints of various kinds as discussed in Chapter 4). We need to honour this process rather than complain about bureaucracy and regulation.

One of the paradoxes of the rural agricultural economy is the idea that we should be 'adding value' to crops. In the case of food, adding economic value usually has two less desirable effects. First, it means processing further, in other words contradicting the human health benefits of eating less processed food, and second, adding an industrial process to the environmental impact, increasing the carbon and other footprints. Could this effort go instead into enhancing trust in the less processed food, via the bureaucratic processes of quality control and standards implementation? Is this a different and maybe better, albeit more hidden, kind of 'adding value'? It would be about the quality of the system of provision, the assurance of different types of environmental and human health 'goods'. The suggestion that we might best add value to food for mass consumption in less visible ways is challenging to the current economic mind-set, which might be uneasy about paying for something as elusive as trust. It is also challenging to that version of romancing local food that is more concerned with the beautiful end product at the farmer's market than the tedious office processes by which it got there and can be verified. But just as harmful old drought might have productive agency, so boring old computer work might add more value than using wheat to create better ice cream. Or at least, they both provide us with tools to think these issues through.

None of this is easy, of course. By the end of 2010, what should have been an excellent harvest had been in many places destroyed by rain or flood. In Victoria a plague of locusts descended, eating their way into the suburban gardens of Melbourne. Throughout the eastern wheat belt mice bred in their millions, munching through wheat in the paddock, as well as harvested wheat in storage. By the spring of 2011 it was reported that mice had damaged 50 to 70 per cent of some crops in southern NSW. In the supermarket wars, one of the Big Supermarkets dropped the price of its generic white sliced loaf to A$1 and two litres of milk to A$2, invoking the ire of farmers, who claimed these prices make agriculture unsustainable, and a mixed response from shoppers.

***

In a chain of connection with distant ancestors, billions of people build deceptively simple wheat based staples like bread and pasta into the daily diets of themselves and their children. They may or may not be conscious of the plantiness that has made this possible for many thousands of years, or its current expression in landscapes such as the NSW wheat belt. In tracing some of the layered processes through which wheat and people have become ingrained in one another's lives, we have also tried to create openings for new kinds of conversations about the future. If the conceptualisation and practice of agriculture is so deeply implicated in the culture/nature binary in Western thought, it is possible that rethinking agriculture – and finding better ways to practise it – will be an important part of reconfiguring the damaging ontologies of modernity.

Appendix 1 Farming Decision Matrix

	Post-Harvest	Pre-Sowing	Sowing	Early In-crop	Mid-Crop	Late Crop	Harvest
Overall strategy	Reflection and review. What did/ didn't go well and why? Analysis of previous yield maps	Choose and source seed varieties; consider expense vs ease of sourcing. Prepare seed	Wait for break; sow dry or wet? If it rains early – is this the break? Will there be another break? Finalise what to sow and where	Monitor crop: for germination, pests and diseases	Monitor crop for maturation; disease, pests, frost and rain. Manage stock on grazing wheats	Watch for late frosts, thunderstorms, hail, pests and diseases	Harvest and store or move grain as quickly and efficiently as possible
Chemical decisions		Purchasing and application of pre-emergent chemicals; clean seed		Top dressing with urea	Spraying as needed. Weekly visit by agronomist	Spraying as needed. Weekly visit of agronomist	
Financial decisions		Consider and purchase insurance	Consider forward selling this crop; consider forward selling next year's crop	Estimate yields; predict profit/loss			Marketing crop if not pre-sold
Decisions about wheat	Consider previous performance of varieties	Consider variety options; long and/or short season, disease resistance, grazing. Buy, prepare/clean seed	Sow treated seed; sowing choices dependant on timing of the break	Monitor plant development; looking for growth stages and monitoring for disease	Monitor plant development	Monitor grain development, seed filling. Estimating yield and quality	Recording yield and quality

Appendix 1 Farming Decision Matrix 9 (continued)

	Post-Harvest	Pre-Sowing	Sowing	Early In-crop	Mid-Crop	Late Crop	Harvest
Farming practises and decisions	Tillage and stubble management; paddock allocation	Spray and clean paddocks as required. Prepare and maintain irrigation infrastructure	Sowing		Manage grazing as appropriate; stock in/out of crop. Spraying as needed	Spraying as needed.	Harvesting, storage and grain segregation logistics
Machinery decisions	Review own machinery vs. lease, or use contractors	Ordering spare parts, maintenance of machinery as required				Maintenance of header, storage facilities, ordering of spare parts	Sharing machinery, helping on other properties
Crucial nature events to monitor		Hope for early rain to germinate weed seeds for pre-crop herbicide application	Growing season; need rain	Growing season; need rain	Heads starting to fill; need rain but no frost or really hot days	Grains maturing; need dry conditions, no storms, insects, pests	Need dry weather, no rain or storms

Appendix 2 Experiences of Drought, December 2006 – May 2007

Similarities to normal droughts	Farmer/s
Every year is different I don't doubt that there is some climate change, with what we're putting into the atmosphere but the drought hasn't been a climate ... change thing, it's been a drought, it's been a pendulum effect. I don't know whether you call it normal or not. It's things that you learn to live with. Normally in bad years you'll have less commodity but it will be higher value We've got no say over the drought We are just going through a cycle at the moment	A D (older*) F (older) G M R

Differences with normal droughts	
Worst drought on our records and they go back to 1882 I suppose there hasn't been a lot of happy moments in the last six or seven years. But before that it used to just amaze me, like to see a beautiful stand of wheat and it was green and healthy and just amazing Worst from a production point of view that anyone can remember It's a lot, lot drier than I've ever seen it Three years prior to this year, the three driest autumns on record, since records have been kept in the district since 1884 ... these droughts are worse than the droughts that were there when I started farming because I believe with those droughts and the sophisticated farming methods we have now, we would have grown a lot of wheat in some of those droughts Worse than the 1940s because so prolonged Historically a lot of the droughts seem to have run for sort of one or two years and then you might get a good year and then sort of another year or two, but this one seems to have been a lot more consistent. Would that be a fair comment K? Yes Back in the 1940s ... it was very dry years ... but not as bad as this When you strike the worst drought in 100 years it really does test you	H (older) N O P (older) Q (older) S (older) T and K (older) V (older) W

Changed practices because of this drought	
The drought has been the biggest driver in change We've cut most of it for hay which we've never done before This is probably the first time I've ever been short of seed [in a drought] There's no way you can store grain and stuff ... like five years possibly going to six ... [we've stopped planting trees in the last five years] because there's been no moisture and it would have been just incredibly disappointing to ... see them all die	C O P (older) R T

Note: *'Older' includes here those participants who are in their sixties and/or retired from farming.

Bibliography

(ABARE) Australian Bureau of Agriculture and Resource Economics. 2006. *Australian Commodity Statistics 2006*. Canberra: Commonwealth of Australia, 367.

(ABARE) Australian Bureau of Agriculture and Resource Economics. 2009. *Australian Crop Report, February 09*. Canberra: Commonwealth of Australia, 149, 24.

(ABARE) Australian Bureau of Agriculture and Resource Economics. 2010. *Australian grains: Financial performance of grains producing farms, 2007–08 to 2009–10, 10.1*. Canberra: Commonwealth of Australia.

(ABARES) Australian Bureau of Agricultural and Resource Economics and Sciences. 2010. *Australian Commodity Statistics 2010*. Canberra: Commonwealth of Australia.

(ABC) Australian Broadcasting Corporation. 2008. *Gunnedah declared natural disaster area*. Available at http://abcscience.com.au/news/2008-11-30/gunnedah-declared-natural-disaster-area/223994 [accessed: 20 October 2011].

(ABC) Australian Broadcasting Corporation. 2009. *Flour Power*. Landline Program. Available at http://www.abc.net.au/landline/content/2008/s2545729.htm [accessed: 21 August 2011].

(ABS) Australian Bureau of Statistics. (n.d.) *Rural Industries and Settlement and Meteorology*. Australian Bureau of Statistics. ABS Cat. No. 7101.0. Canberra: ABS.

(ABS) Australian Bureau of Statistics. 1999. *National Nutrition Survey, Foods Eaten Australia, 1995*, ABS Cat. No. 4804.0. Canberra: ABS.

(ABS) Australian Bureau of Statistics. 2000. *Apparent Consumption of Food Stuffs 1997–98 and 1998–99*. ABS Cat. No. 4306.0. Canberra: ABS.

(ABS) Australian Bureau of Statistics. 2006. *Year Book Australia, 2006*. ABS Cat. No. 1301.0. Canberra: ABS.

(ABS) Australian Bureau of Statistics. 2008a. *Principal Agricultural Commodities, Australia, Preliminary, 2007–2008*, ABS Cat. No. 7111.0. Canberra, ABS.

(ABS) Australian Bureau of Statistics. 2008b. *Historical Selected Agricultural Commodities by State (1861–Present), 2007*, ABS Cat. No. 7124.0. Canberra, ABS.

(ABS) Australian Bureau of Statistics. 2010. *2009–2010 Year Book Australia*, ABS Cat no. 1301.0. Canberra, ABS.

(ABS) Australian Bureau of Statistics. 2011. *Agricultural Commodities 2009–2010*, ABS Cat No. 7121.0. Canberra, ABS.

(ACCC) Australian Competition and Consumer Commission. 1999. Submission to the Joint Select Committee on the Retailing Sector, Dickson.

Acosta-Michlik, L., Kelkar, U. and Sharma, U. 2008. A critical overview: Local evidence on vulnerabilities and adaptations to global environmental change in developing countries. *Global Environmental Change* 18(4), 539–542.

Adams, M. 2008. Foundational myths: Country and conservation in Australia. *Transforming Cultures eJournal* 3 (1). Available at http://epress.lib.uts.edu.au/ojs/index.php/TfC/article/view/684.

Adger, W.N., Huq, S., Brown, K., Conway, D. and Hulme, M. 2003. Adaptation to climate change in the developing world. *Progress in Development Studies* 3, 179–19.

Alcamo, J., Dronin, N., Endejan, M., Golubev, G. and Kirilenko, A. 2007. A new assessment of climate change impacts on food production shortfalls and water availability in Russia. *Global Environmental Change* 17(3–4), 429–444.

Alston, M. 1995. Women and their work on Australian farms. *Rural Sociology* 60, 521–32.

Anderson, K. 1997. A walk on the wild side: A critical geography of domestication. *Progress in Human Geography* 21(4), 463–485.

Anderson, K. 2003. White natures: Sydney's royal agricultural show in posthumanist perspective. *Transactions of the Institute of British Geographers* (NS) 28, 422–441.

Anderson, K. 2005. Griffith Taylor lecture, Geographical Society of New South Wales, 2004: Australia and the 'State of Nature/Native'. *Australian Geographer* 36(3), 267–282.

Anderson, K. 2008. *Race and the Crisis of Humanism.* London: Routledge.

Andree, P., Dibden, J., Higgins, V. and Cocklin, C. 2007. Contesting Productivism? Alternative Agri-Food Networks in Australia. Submission to Working Group 21: Local Food, Identity, and Rural Sustainable Development. XXII Congress of the European Society for Rural Sociology; Wageningen, 20–24 August 2007. Mobilities, Vulnerabilities and Sustainabilities: New questions and challenges for rural Europe.

Andree, P., Dibden, J., Higgins, V. and Cocklin, C. 2010. Competitive productivism and Australia's emerging 'alternative' agri-food networks: Producing for farmers; markets in Victoria and beyond. *Australian Geographer* 41(3), 307–322.

Antony, G. and Brennan, J.P. 1988. Improvements in yield potential and bread-making characteristics of wheat in NSW: 1925–26 to 1983–84. *Miscellaneous Bulletin* 55, Agdex 112/820, NSW Agriculture and Fisheries.

(ASX) Australian Securities Exchange. 2011. *Products. Australian Milling Wheat.* Available at http://www.asx.com.au/products/grain-contract-specifications-australian-milling-wheat.htm [accessed: 10 October 2011].

Atchison, J. 2009. Human impacts on *Persoonia falcata*. Perspectives on post-contact vegetation change in the Keep River region, Australia, from contemporary vegetation surveys. *Vegetation History and Archaeobotany* 18(2), 147–157.

Atchison, J., Head, L. and Fullagar, R. 2005. Archaeobotany of fruit seed processing in a monsoon savanna environment: Evidence from the Keep River Region, Northern Territory, Australia. *Journal of Archaeological Science* 32(2), 167–181.

Atchison, J., Head, L. and Gates, A. 2010. Wheat as food, wheat as industrial substance; Comparative geographies of transformation and mobility. *Geoforum* 41(2), 236–246.

Atkins, P. 2010. *Liquid Materialities: A History of Milk, Science and the Law.* Surrey, England: Ashgate.

Barker, K. 2008. Flexible boundaries in biosecurity: Accommodating gorse in *Aotearoa* New Zealand. *Environment and Planning A* 40, 1598–1614.

Barkworth, M.E. 2000. Changing perceptions of the Triticeae, in *Grasses: Systematics and Evolution,* edited by S. Jacobs and J. Everett. Collingwood: CSIRO, 110–122.

Barrett, C. and Maxwell, D. 2005. *Food Aid After Fifty Years. Recasting its Role.* London: Routledge.

Basher, R.E., Pittock, A.B., Bates, B., Done, T., Gifford, R.M., Howden, S.M., Sutherst, R., Warrick, R., Whetton, P., Whitehead, D., Williams, J.E. and Woodward, A. 1998. Australasia, in *The Regional Impacts of Climate Change. An Assessment of Vulnerability,* edited by R.T. Watson, Z.M.C. and R.H. Moss. Cambridge: Cambridge University Press, 105–148.

Bear, C. and Eden, S. 2011. Thinking like a fish? Engaging with nonhuman difference through recreational angling. *Environment and Planning D: Society and Space* 29, 336–52.

Beisel, U. 2010. Jumping hurdles with mosquitoes? *Environment and Planning D: Society and Space* 28, 46–49.

Bell, D. and Valentine, G. 1997. *Consuming Geographies. We Are Where We Eat.* London: Routledge.

Belt, R. and Sheridan, H. 2008. Flooding rain beats down on Tamworth. *The Sydney Morning Herald,* 30 November. Available at http://www.smh.com.au/news/national/flooding-rains-beat-down-on-tamworth/2008/11/29/1227491892367. html [accessed: 20 October 2011].

Bennett, M. 2005. A long time working: Aboriginal labour on the Coolangatta Estate, 1822–1901. Paper presented to 'The Past is Before Us', History Cooperative University of Sydney, 2005. Available at http://www.historycooperative.org/proceedings/asslh/bennett.html [accessed: 17 October 2011].

Benicci, A. 2008. Origin and evolution of land plants. *Communicative and Integrative Biology* 1(2), 212–218.

Berghoff, W. 1998. Längsschnitt durch ein Getreidekorn. aid infodienst. Bonn: Verbraucherschutz Ernährung.

Bishop, P. 1991. Constable country: Diet, landscape and national identity. *Landscape Research* 16(2), 31–36.

Bissell, D. 2010. Vibrating materialities: Mobility-boy-technology relations. *Area* 42(4), 479–486.

Biswas, W.K., Barton, L. and Carter, D. 2008. Global warming potential of wheat production in Western Australia: A life cycle assessment. *Water and Environment Journal* 22(3), 206–216.

Blunt, A. 2007. Cultural geographies of migration: Mobility, transnationality and diaspora. *Progress in Human Geography* 31(5), 684–694.

Bobe, R. and Behrensmeyer, A.K. 2004. The expansion of grassland ecosystems in Africa in relation to mammalian evolution and the origin of the genus *Homo*. *Palaeogeography, Palaeoclimatology, Palaeoecology* 207(3–4), 399–420.

(BOM) Bureau of Meteorology. 2005. Australian average rainfall map, Annual. Available at http://www.bom.gov.au/jsp/ncc/climate_averages/rainfall/index. jsp [accessed: 26 February 2009].

Botkin, D.B. and Beveridge, C.E. 1997. Cities as environments. *Urban Ecosystems* 1(1), 3–19.

Botterill, L.C. 2007. Doing it for the Growers in Iraq?: The AWB, Oil-for-Food and the Cole Inquiry. *The Australian Journal of Public Administration* 66(1), 4–12.

Botterill, L.C. 2011. Life and death of an institution: The case of collective wheat marketing in Australia. *Public Administration* 89, 629–643.

Botterill, L.C. and McNaughton, A. 2008. Laying the foundations for the wheat scandal: UN sanctions, private actors and the Cole Inquiry. *Australian Journal of Political Science* 43(4), 583–598.

Braidotti, G. 2011. Yesterday's aid recipient is today's R&D partner. *Partners in Research for Development* June-August 2011, 9–11. Canberra: Australian Centre for International Agricultural Research. Available at http://aciar.gov.au [accessed: 16 October 2011].

Bradshaw, B., Dolan, H. and Smit, B. 2004. Farm-level adaptation to climatic variability and change; crop diversification in the Canadian prairies. *Climatic Change* 67(1), 119–141.

Bradshaw, M. and Weaver, R. 1993. *Physical Geography. An Introduction to Earth Environments.* St. Louis: Mosby-Year Book.

(BRI) BRI Australia Ltd. 2003. *The Australian Baking Industry. A Profile.* Canberra: Commonwealth of Australia.

(BRI) Bread Research Institute of Australia. 1989. *Australian Bread Making Handbook.* Kensington, NSW: TAFE Educational Books in conjunction with (BRI) The Bread Research Institute of Australia.

Broehl, W.G. 1992. *Cargill. Trading the World's Grain.* Hanover: University Press of New England.

Brooks, N., Adger, W.N. and Kelly, P.M. 2005. The determinants of vulnerability and adaptive capacity at the national level and the implications for adaptation. *Global Environmental Change* 15(2), 151–63.

Bryant, C.R., Smith, B., Brklacich, M., Johsnton, T.R., Smithers, J., Chiotti, Q. and Singh, B. 2000. Adaptation in Canadian agriculture to climatic variability and change. *Climatic Change* 45, 181–201.

Bryant, L. and Pini, B. 2011. *Gender and Rurality.* New York and London: Routledge.

Burton, I. and Lim, B. 2005. Achieving adequate adaptation in agriculture. *Climatic Change* 70(1–2), 191–200.

Calderini, D.F. and Slafer, G.A . 1998. Changes in yield and yield stability in wheat during the 20th century. *Field Crops Research* 57, 335–347.

Caligari, P.D.S. and Brandham, P.E. (eds). 2001. *Wheat Taxonomy: the Legacy of John Percival*, edited by P.D.S. Caligari and P.E. Brandham. London: Linnean Society, Linnean Special Issue 3.

Callaghan, A.R. and Millington, A.J. 1956. *The Wheat Industry in Australia*. Sydney: Angus and Robertson.

Campbell, W.A. 1911. An historical sketch of William Farrer's work in connection with his improvements in wheats for Australian conditions. *Proceedings of the Australian Association for the Advancement of Science* Section G2, 13, 525–536.

Carberry, P., Keating, B., Bruce, S. and Walcott, J. 2010. Technological innovation and productivity in dryland agriculture in Australia. Canberra: a joint paper prepared by ABARE-BRS and CSIRO.

Cassidy, R. and Mullin, M. (eds). 2007. *Where The Wild Things Are Now: Domestication Reconsidered.* Oxford, UK: Berg.

Castree, N. 2002. False antitheses? Marxism, nature and actor networks. *Antipode* 34(1), 111–46.

Cavalier-Smith, T. 1998. A revised six-kingdom system of life. *Biological Reviews* 73(3), 203–266.

Chase, A.K. 1989. Domestication and domiculture in northern Australia: A social perspective, in *Foraging and Farming: The Evolution of Plant Exploitation*, edited by D.R. Harris and G.C. Hillman, London: Unwin Hyman, 42–54.

Clark, N. 2011. *Inhuman Nature: Sociable Life on a Dynamic Planet.* London: Sage.

Cloke, P. and Pawson, E. 2008. Memorial trees and treescape memories. *Environment and Planning D: Society and Space* 26, 107–22.

Cocklin, C. and Dibden, J. 2009. Systems in peril: Climate change, agriculture and biodiversity in Australia. *IOP Conference Series: Earth and Environmental Science* 8, 1–21.

Coghlan, T.A. 1901. *Statistics: Six States of Australia and New Zealand 1861–1900.* Sydney: William Applegate Gullick.

Cook, I. 2004. Follow the thing: Papaya. *Antipode* 36(4), 642–664.

Cook, I., et al. 2006. Geographies of Food: Following. *Progress in Human Geography* 30(5), 655–666.

Commonwealth of Australia. 2008. *Australian Food Statistics 2007.* Canberra: Australian Government Department of Agriculture, Fisheries and Forestry.

Cordell, D., Drangert, J. and White, S. 2009. The story of phosphorus: Global food security and food for thought. *Global Environmental Change* 19(2), 292–305.

Cornell, H.J. and Hoveling, A.W. 1998. *Wheat Chemistry and Utilization.* Basel: Technomic.

Cosgrove, D.J. 2005. Growth of the plant cell wall. *Nature* Reviews, *Molecular Cell Biology* 6, 850–861.

Coulthard, S. 2008. Adapting to environmental change in artisanal fisheries – Insights from a South Indian Lagoon. *Global Environmental Change* 18(3), 479–489.

Coursey, D.G. 1973. Hominid evolution and hypogeous plant foods. *Man* 8, 634–635.

Crang, M. 2003. Qualitative methods: Touchy, feely, look-see? *Progress in Human Geography* 27(4), 494–504.

Cresswell, T. 2006. *On the Move: Mobility in the Modern Western World*. London: Routledge.

Cresswell, T. 2010. Towards a politics of mobility. *Environment and Planning D: Society and Space* 28, 17–31.

Cronon, W. 1991. *Nature's Metropolis. Chicago and the Great West*. New York: W.W. Norton and Company.

Crosby, A.W. 1972. *The Columbian Exchange: The Biological and Cultural Consequences of 1492*. Westwood, CO: Greenwood Press.

Crosby, A.W. 1986. *Ecological Imperialism: The Biological Expansion of Europe, 900–1900*. Cambridge, UK: Cambridge University Press.

(DAFF) Australian Government Department of Agriculture, Fisheries and Forestry. 2009. *Drought assistance and Exceptional circumstances*. Available at http://www.daff.gov.au/agriculture-food/drought/ec/background [accessed: 14 October 2009].

Davidson, B. 1987. Agriculture, in *Australian Historical Statistics*, edited by W. Vamplew. Sydney: Fairfax Syme & Welden Associates.

Davis, M.A., Chew, M.K., Hobbs, R.J., Lugo, A.E., Ewel, J.J., Vermeij, G.J., Brown, J.H., Rosenzweig, M.L., Gardener, M.R., Carroll, S.P., Thompson, K., Pickett, S.T.A., Stromberg, J.C., Del Tridici, P., Suding, K.N., Ehrenfeld, J.G., Grime, J.P., Mascaro, J. and Briggs, J.C. 2011. Don't judge species on their origins. *Nature* 474, 153–4.

Denham, T. 2007a. Early to Mid-Holocene plant exploitation in New Guinea: Towards a contingent interpretation of agriculture in *Rethinking Agriculture. Archaeological and Ethnoarchaeological Perspectives,* edited by T. Denham, J. Iriarte and L. Vrydaghs. California: Left Coast Press, 78–108.

Denham, T. 2007b. Early agriculture. Recent conceptual and methodological developments, in *The Emergence of Agriculture, A Global View*, edited by T. Denham and P. White. London: Routledge, 1–25.

Denham, T. and White, P. (eds). 2007. *The Emergence of Agriculture, A Global View.* London: Routledge.

Denham, T., Fullagar, R. and Head, L. 2009. Plant exploitation on Sahul: From colonisation to the emergence of regional specialisation during the Holocene. *Quaternary International* 202(1–2), 29–40.

Derrick, J.W. and Dumaresq, D.C. 1999. Soil chemical properties under organic and conventional management in southern New South Wales. *Australian Journal of Soil Research* 37, 1047–1055.

Dessai, S., Adger, W.N., Hulme, M., Turnpenny, J., Kohler, J. and Warren, R. 2004. Defining and experiencing dangerous climate change – An editorial essay. *Climatic Change* 64(1–2), 11–25.

Dibden, J. and Cocklin, C. (eds) 2005. *Sustainability and Change in Rural Australia*. Sydney: University of New South Wales Press.

Dibden, J. and Cocklin, C. 2009. 'Multifunctionality': Trade protectionism or a new way forward? *Environment and Planning A* 41: 163-182.

Dibden, J. and Cocklin, C. 2010. Re-mapping regulatory space: The new governance of Australian dairying. *Geoforum* 41(3), 410–422.

Dixon, J., Braun, B., Kosina, P. and Crouch, J. (eds) 2009. *Wheat Facts and Futures 2009*. Mexico: CIMMYT.

Donald, C.M. 1963. Grass or crop in the land use of tomorrow. *Australian Journal of Science* 25, 386–395.

Dunn, K.M. 2009. Meinig, in *International Encyclopedia of Human Geography*, Volume 7, edited by R. Kitchin and N. Thrift. Oxford: Elsevier, 48–50.

Dunsdorfs, E. 1956. *The Australian Wheat-Growing Industry 1788–1948*. Carlton, Melbourne: University Press.

(DPI) Department Primary Industries and Agriculture, NSW. 2010. Winter Crop Variety Sowing Guide. Available at http://www.grdc.com.au/director/events/grdcpublications/sowingguides [accessed: 22 October 2011].

Edensor, T. 2010. Introduction: Thinking about rhythm and space, in *Geographies of Rhythm: Nature, Place, Mobilities and Bodies,* edited by T. Edensor. Farnham: Ashgate, 1–20.

Ellis E.C. and Ramankutty N. 2008. Putting people in the map: Anthropogenic biomes of the world. *Frontiers in Ecology and the Environment* 6(8), 439–447.

Emel, J., Wilbert, C. and Wolch, J. 2002. Animal geographies. *Society and Animals* 10(4), 407–412.

Ennos, R. and Sheffield, E. 2000. *Plant Life*. Manchester: Blackwell Science.

Evers, A.D. and Bechtel D.B. 1998. Microscopic structure of the wheat grain, in *Wheat: Chemistry and Technology*, Volume 1, edited by Y. Pomeranz. St Paul, Minnesota: American Association of Cereal Chemists, Inc., 47–95.

Fairweather, H. and Cowie, A. 2007. Climate change research priorities for NSW primary industries, discussion paper, NSW Department of Primary Industries.

(FAO) Food and Agriculture Organisation of the United Nations. 2009. *The State of Food Insecurity in the World. Economic Crises – Impacts and Lessons Learned*. Rome: FAO.

(FAOSTATS) Food and Agriculture Organisation of the United Nations, Statistics Division. 2011. *Production and Trade Data*. Available at http://faostat.fao.org/site/339/default.aspx [accessed: 2 September 2011].

Feldman, M. 2001. Origin of Cultivated Wheat, in *The World Wheat Book. A History of Wheat Breeding*, edited by A.P. Bonjean and J.F. Angus. Paris: Lavoisier, 3–56.

Fennema, O. (ed.). 1985. *Food Chemistry*, 2nd Edition. New York: M., Dekker Inc.

Fincher, R. and Iveson, K. 2008. *Planning and Diversity in the City: Redistribution, Recognition and Encounter*. New York: Palgrave Macmillan.

Firn, R. 2004. Plant intelligence: An alternative point of view. *Annals of Botany* 93(4), 345–351.

Fischer, J., Stott, J. and Law, B. 2010. The disproportionate value of scattered trees. *Biological Conservation* 143, 1564–1067.

Flugge, T. 1997. *Innovation in Agricultural Product Marketing – The Australian Wheat Industries Experience. Farrer Memorial Oration, 1997*. Available at http://www.dpi.nsw.gov.au/__data/assets/pdf_file/0018/270252/1997-innovations-in-agriculture-product-marketing-the-australian-wheat-industry-experience-Trevor-Flugge.pdf [accessed: 16 October 2011].

Franklin, A. 2006. Burning cities: A posthumanist account of Australians and eucalypts. *Environment and Planning* D 24, 555–576.

(FSANZ) Food Standards Australia New Zealand. 2009a. *Australia New Zealand Food Standards Code*. Available at http://www.foodstandards.gov.au/thecode/foodstandardscode/ [accessed: 6 August 2009].

(FSANZ) Food Standards Australia New Zealand. 2009b. *Confectionery recall due to undeclared gluten allergen*. Available at http://www.foodstandards.gov.au/foodmatters/foodrecalls/archiveconsumerlevelrecalls/confectionerylabelli4334.cfm[accessed: 6 August 2009].

(FSANZ) Food Standards Australia New Zealand. 2009c. *Wheat: Ingredients to avoid if you are allergic to wheat*. Available at http://www.foodstandards.gov.au/foodmatters/foodallergies/wheatallergy.cfm [accessed: 6 August 2009].

Ford, J.D., Smit, B., Wandel, J., Allurut, M., Shappa, K., Ittusarjuat, H. and Qrunnut, K. 2008. Climate change in the Arctic: Current and future vulnerability in two Inuit communities in Canada. *Geographical Journal* 174(1), 45–62.

Fullagar, R., Field, J. and Kealhofer, L. 2008. Grinding stones and seeds of change: Starch and phytoliths as evidence of plant food processing, in *New Approaches to Old Stones: Recent Studies of Ground Stone Artifacts*, edited by Y.M. Rowan and J.R. Ebeling. London: Equinox Press, 159-172.

Gandy, M. 2002. *Concrete and Clay: Reworking Nature in New York*. Cambridge MA: MIT Press.

Gates, A. 2009. *Riverina Flour Mills Research Project: Final Report to the Royal Australian Historical Society*. Wollongong: University of Wollongong.

Geissler, P.W. and Prince, R.J. 2009. Plants, bodies, minds and cultures in the work of Kenyan ethnobotanical knowledge. *Social Studies of Science* 39(4), 599–634.

Gibbons, P., Lindenmayer, D.B., Fischer, J., Manning, A., Weinberg, A., Seddon, J., Ryan, P. and Barret, G. 2008. The future of scattered trees in agricultural landscapes. *Conservation Biology* 22, 1309–1319.

Gibson, C., Head, L., Gill, N. and Waitt, G. 2011. Climate change and household dynamics: Beyond consumption, unbounding sustainability. *Transactions, Institute of British Geographers* 36(1), 3–8.

Gill, N. and Paterson, A. 2007. A work in progress: Aboriginal people and pastoral cultural heritage in Australia, in *Geographies of Australian Heritages: Loving a Sunburnt Country?* edited by R. Jones and B.J. Shaw. Aldershot: Ashgate, 113–132.

Gilmore, R. 1982. *A Poor Harvest. The Clash of Policies and Interests in the Grain Trade.* New York: Longman.

Ginn, F. 2008. Extension, subversion, containment: Eco-nationalism and (post) colonial nature in Aotearoa New Zealand. *Transactions of the Institute of British Geographers* 33, 335–353.

Ghosh, J. 2010. The unnatural coupling: Food and global finance. *Journal of Agrarian Change* 10(1), 72–86.

(GRDC) Grains Research and Development Corporation and (GCA) Grains Council of Australia. 2004. *Directory of Linkages in the Australian Grains Industry as at February 2004.* Prepared by GRDC and GCA as a background document to the Grains Industry Strategic Plan; Kronos Corporate 2004.

(GTA) Grain Trade Australia. 2011a. *Section 2 – Wheat Standards 2011/12 Season.* Available at http://www.graintrade.org.au/commodity_standards [accessed: 19 August 2011].

(GTA) Grain Trade Australia. 2011b. *Wheat Receival Standards 2010–11. Wheat Chart.* Available at http://www.graincorp.com.au/prodserv/SL/Pages/receivalstandards.aspx [accessed: 19 August 2011].

Goodall, H. 1996. *Invasion to Embassy: Land in Aboriginal Politics in New South Wales, 1770–1972.* Sydney: Allen & Unwin.

Goodman, D. 1999. Agro-food studies in the 'age of ecology': Nature, corporeality, bio-politics. *Sociologia Ruralis* 39(1), 17–38.

Gorman-Murray, A. 2010. An Australian feeling for snow: Towards understanding cultural and emotional dimensions of climate change. *Cultural Studies Review* 16 (1), 60–81.

Goswami, O. 1982. Collaboration and conflict: European and Indian capitalists and the jute economy of Bengal, 1919–39. *Indian Economic Social History Review* 19(2), 141–179.

Gott, B. 1982. Ecology of root use by the Aborigines of southern Australia. *Archaeology in Oceania* 17(1), 59–67.

Gott, B. 1983. *Microseris scapigera*: A study of a staple food of Victorian Aborigines. *Australian Aboriginal Studies* 2, 2–18.

Graham, J.B., Dudley, R., Aguilar, N.M. and Gans, C. 1995. Oxygenation of the late Paleozoic oxygen pulse for physiology and evolution. *Nature* 375, 117–120.

Graham, L.E. 1996. Green algae to land plants: An evolutionary transition mini review. *Journal of Plant Research* 109(3), 241–251.

Green, P. and Jabri, B. 2003. Coeliac disease. *The Lancet* 362(9393), 1418-1419.

Gunasekera, D., Kim, Y., Tulloh, C. and Ford, M. 2007. Climate Change. Impacts on Australian Agriculture. *Australian Commodities* 14(4), 657–676.

Hall, M. 2011. *Plants as Persons. A Philosophical Botany.* Albany: State University of New York Press.

Hallam, S.J. 1989. Plant usage and management in southwest Australian Aboriginal societies, in *Foraging and Farming: The Evolution of Plant Exploitation*, edited by D.R. Harris and G.C. Hillman. London: Unwin Hyman, 136–151.

Hamblin, A. and Kyneur, G. 1993. *Trends in Wheat Yields and Soil Fertility in Australia*. Canberra, Australian Government Publishing Service.

Hannah, M. 2002. Aboriginal Workers in the Australian Agricultural Company, 1824–1857. *Labour History* 82, 17–33.

Haraway, D. 2008. *When Species Meet.* Minneapolis: University of Minnesota Press.

Hare, R. 2001. *Durum Wheat in Australia – Past, Present and Future.* Farrer Memorial Oration 2001. Available at http://www.dpi.nsw.gov.au/agriculture/field/field-crops/farrer-memorial-trust/orations/farrer-memorial-oration-2001 [accessed: 16 October 2011].

Harlan, J.R. 1981. The early history of wheat: Earliest traces to the sack of Rome, in *Wheat Science – Today and Tomorrow*, edited by L.T. Evans and W.J. Peacock. Cambridge: Cambridge University Press, 1–19.

Harle, K.J., Howden, S.M., Hunt L.P. and Dunlop, M. 2007. The potential impact of climate change on the Australian wool industry by 2030. *Agricultural Systems* 93(1–3), 61–89.

Harris, D.R. 1977. Alternative pathways toward agriculture, in *Origins of Agriculture,* edited by C.A. Reed. The Hague: Mouton, 179–243.

Harris, D.R. 1989. An evolutionary continuum of people-plant interaction, in *Foraging and Farming: The Evolution of Plant Exploitation*, edited by D.R. Harris, and G.C. Hillman. London: Unwin Hyman, 11–26.

Harris, D.R. 2007. Agriculture, cultivation and domestication: Exploring the conceptual framework of early food production, in *Rethinking Agriculture: Archaeological and Ethnoarchaeological Perspectives*, edited by T. Denham, J. Iriarte, and L. Vrydaghs. California: Left Coast Press Inc, 16–35.

Harvey, G. 2005. *Animism: Respecting the Living World.* London: Hurst & Company.

Hatfield-Dodds, S., Carwardine, J., Dunlop, M., Graham, P., and Klein, C. 2007. Rural Australia Providing Climate Solutions. Preliminary report to the Australian Agricultural Alliance on Climate Change. Canberra: CSIRO Sustainable Ecosystems.

Head, L. 2000. *Second Nature. The History and Implications of Australia as Aboriginal Landscape.* Syracuse: Syracuse University Press.

Head, L., Atchison, J. and Fullagar, R. 2002. Country and garden: Ethnobotany, archaeobotany and Aboriginal landscapes near the Keep River, northwestern Australia. *Journal of Social Archaeology* 2(2), 173–196.

Head, L. and Atchison, J. 2009. Cultural ecology: Emerging human-plant geographies. *Progress in Human Geography* 33(2), 236–245.

Head, L., Atchison, J., Gates, A. and Muir, P. 2011. A fine-grained study of the experience of drought, risk and climate change among Australian wheat farming households. *Annals of the Association of American Geographers* 101(5), 1089–1108.

Head, L., Atchison, J. and Phillips, C. 2011. Rethinking human-plant worlds: Materiality, agency and encounter. Paper presented to the Institute of Australian Geographers Conference, Wollongong.

Head, L. and Muir, P. 2007. *Backyard. Nature and Culture in Suburban Australia.* Wollongong: University of Wollongong Press.

Heathcote, R.A. 1963. Bread or Cake? A geographer and a historian on the nineteenth century wheat frontier: A review. *Economic Geography* 39(3), 176–182.

Hedges, S.B. 2002. The origin and evolution of model organisms. *Nature Reviews* 3, 838–849.

Helfe, S.L. 2001. Hidden food allergies. *Current Opinion in Allergy and Clinical Immunology* 1, 269–271.

Hennessy, K., Fawcett, R., Kirono, D., Mpelsaoka, F., Jones, D., Bathols, J., Whetton, P., Stafford Smith, M., Howden, M., Mitchell, C. and Plummer, N. 2008 *An assessment of the impact of climate change on the nature and frequency of exceptional climatic events.* Canberra: CSIRO and Bureau of Meteorology.

Henrick, P. and Crane, P. 1997. The origin and evolution of plants on land *Nature* 389, 33–39. Republished in *Shaking the Tree Readings from Nature in the History of Life,* edited by H. Gee, 2000. USA: University of Chicago Press.

Hetherington, K. 2003. Spatial textures: Place, touch and praesentia. *Environment and Planning A* 35, 1933–1944.

Heyhoe, E., Kim, Y., Kokic, P., Levantis, C., Ahammad, H., Schneider, K., Crimp, S., Nelson, R., Flood, N. and Carter, J. 2007. Adapting to climate change. Issues and challenges in the agriculture sector. *Australian Commodities* 14(1), 167–178.

Heynen, N., Kaika, M. and Swyngedouw, E. (eds) 2006. *In the Nature of Cities. Urban political ecology and the politics of urban metabolism.* Abingdon: Routledge.

Hird, M. 2009. *The Origins of Sociable Life: Evolution after Science Studies.* Basingstoke: Palgrave Macmillan.

Hitchings, R. 2003. People, plants and performance: On actor network theory and the material pleasures of the private garden. *Social & Cultural Geography* 4(1), 99–114.

Hitchings, R. 2006. Expertise and inability. cultured materials and the reason for some retreating lawns in London. *Journal of Material Culture.* 11(3), 364–381.

Hitchings, R. 2007. How awkward encounters could influence the future form of many gardens. *Transactions of the Institute of British Geographers.* 32(3), 363–376.

Hitchings, R. and Jones, O. 2004. Living with plants and the exploration of botanical encounter within human geographic research practice. *Ethics, Place and Environment* 7(1–2), 3–18.

Hobbs, R.J., Arico, S., Aronson, J., Baron, J.S., Bridgewater, P., Cramer, A.A., Epstein, P.R., Ewel, J.J., Klink, C.A., Lugo, A.E., Norton, D., Ojima, D., Richardson, M., Sanderson, E.W., Valladares, F., Vilá, Zamora, R. and Zobel,

M. 2006. Novel Ecosystems: Theoretical and management aspects of the new ecological world order. *Global Ecology and Biogeography* 15, 1–7.

Hobbs, R.J., Higgs, E. and Harris, J.A. 2009. Novel Ecosystems: Implications for conservation and restoration. *Trends in Ecology and Evolution, Opinion* 24, 599–605.

Hodder, I. 2007. Çatalhöyük in the Context of the Middle Eastern Neolithic. *Annual Review of Anthropology* 36, 105–20.

House, A.P.N., MacLeod, N.D., Cullen, B., Whitbread, A.M., Brown, S.D. and McIvor, J.G. 2008. Integrating production and natural resource management on mixed farms in eastern Australia: The cost of conservation in agricultural landscapes. *Agriculture, Ecosystems and Environment* 127(3–4), 153–165.

Howden, M. and Jones, R. 2001. Costs and benefits of CO_2 increase and climate change on the Australian wheat industry. Canberra: Attorney Generals Office.

Howden, S.M., Soussana, J.F., Tubiello, F.N., Chhetri, N., Dunlop, M. and Meinke, H. 2007. Adapting agriculture to climate change. *Proceedings National Academy of Sciences* 104(5), 19691–19696.

Howdle, P.D., Losowsky, M.S. 1992. Coeliac disease in adults, in *Coeliac Disease*, edited by M.N. Nash. Oxford: Blackwell Scientific Publications, 49–80.

Howitt, R. 1993. 'A world in a grain of sand': Towards a reconceptualisation of geographical scale. *Australian Geographer* 24, 33–44.

Hulme, M. 2008. Geographical work at the boundaries of climate change. *Transactions of the Institute of British Geographers* 33(1), 5–11.

Huttner, E. and Debrand, M. 2001. Contribution of genomics to wheat improvement, in *The World Wheat Book. A History of Wheat Breeding*, edited by A.P. Bonjean and J.F. Angus. Paris: Lavoisier, 1061–1080.

Ingold, T. 2000. *The Perception of the Environment. Essays on Livelihood, Dwelling and Skill*. London: Routledge.

Instone, L. 2010. Encountering native grasslands: Matters of concern in an urban park *Australian Humanities Review* 49. Available at http://www.australianhumanitiesreview.org/archive/Issue-November-2010/instone.html) [accessed: 30 May 2011].

Jeans, D.N. (ed.). 1987. *Australia, A Geography,* Volume 2. Space and Society. Sydney: Sydney University Press.

Johnston, L. 2010. Sites of excess: The spatial politics of touch for drag queens in Aotearoa New Zealand. *Emotion, Space and Society* 5(1), 1–9.

Jones, R. 1975. The Neolithic, Palaeolithic and the Hunting Gardeners: Man and Land in the Antipodes, in *Quaternary Studies. Selected Papers from IX Inqua Congress, Christchurch, N.Z., December 1973*, edited by R.P. Suggate and M.M. Cresswell. Christchurch: Royal Society of New Zealand, 21–34.

Jones, O. and Cloke, P. 2002. *Tree Cultures. The Place of Trees and Trees in Their Place*. Oxford: Berg.

Jones, M. and Brown, T. 2007. Selection, cultivation and reproductive isolation: A reconsideration of the morphological and molecular signals of domestication, in *Rethinking Agriculture. Archaeological and Ethnoarchaeological*

Perspectives, edited by T. Denham, J. Iriarte and L. Vrydaghs. California: Left Coast Press, 36–49.

Kaika, M. 2005. *City of Flows: Modernity, Nature, and the City*. New York: Routledge.

Kearnes, M.B. 2003. Geographies that matter – The rhetorical deployment of physicality? *Social & Cultural Geography* 4(2), 139–152.

Keith, D. 2004. *Ocean Shores to Desert Dunes. The Native Vegetation of New South Wales and the ACT*. Hurstville: Department of Environment and Conservation (NSW).

Kelkar, U., Narula, K.K., Sharma, V.P. and Chandna, U. 2008. Vulnerability and adaptation to climate variability and water stress in Uttarakhand State, India. *Global Environmental Change* 18(4), 564–574.

Kellogg, E.A. 1998. Relationship of cereal crops and other grasses. *Proceeds of the National Academy of Sciences USA* 95, 2005–2010.

Kellogg, E.A. 2001. Evolutionary history of the grasses. *Plant Physiology* 125(3), 1198–1205.

Keyzer, M., Merbis, M., Nube, M. and van Wesenbeek, L. 2008. Food, Feed and Fuel: When competition starts to bite. Amsterdam: OBCentre for World Food Studies.

Kinlan, B. and Gaines, S. 2003. Propagule dispersal in marine and terrestrial organisms: A community perspective. *Ecology* 84(8), 2007–2020.

Kirby, E.J.M. and Appleyard, M. 1987. *Cereal Development Guide*, 2nd Edition. Warwickshire, England: National Agricultural Centre.

Kirkegaard, J.A., Gardner, P.A., Angus, J.F. and Koetz, E. 1994. Effect of Brassica break crops on the growth and yield of wheat. *Australian Journal of Agricultural Research* 45(3), 529–545.

Knobloch, F., 1996. *The Culture of Wilderness. Agriculture as Colonization in the American West*. Chapel Hill: University of North Carolina Press.

Kronos (Kronos Corporate). 2002. A Review of Structural Issues in the Australian Grain Market. Armadale Vic: Kronos Corporate Pty. Ltd.

Kull, C.A. and Rangan, H. 2008. Acacia exchanges: Wattles, thorn trees, and the study of plant movements. *Geoforum* 39(3), 1258–1272.

Ladle, R.J. and Jepson, P. 2008. Towards a biocultural theory of avoided extinction. *Conservation Letters* 1(3), 111–118.

Laris, P. 2011. Humanizing savanna biogeography: Linking human practices with ecological pattern in a frequently burned savanna of southern Mali. *Annals of the Association of American Geographers* 101(5), 1067-1088.

Larsen, J., Urry, J. and Axhausen, K.W. 2006. Mobilities, Networks, Geographies. Aldershot: Ashgate.

Latour, B. 1993. *We Have Never Been Modern*. New York: Harvester Wheatsheaf.

Law, J. 2010. The materials of STS, in *The Oxford Handbook of Material Culture Studies*, edited by D. Hicks and M.C. Beaudry. Oxford: Oxford University Press, 173–188.

Lawrence, L. and Caddick, L. 2006. Popularity of harvest bags heating up. *Farming Ahead* 171, April. Canberra: Commonwealth Scientific Industrial Research Organization (CSRIO).

Leigh, G.J. 2004. *The World's Greatest Fix. A History of Nitrogen and Agriculture.* Oxford: Oxford University Press.

Lenzen, M., Murray, J., Sack, F. and Wiedmann, T. 2007. Shared producer and consumer responsibility – Theory and practice. *Ecological Economics* 61, 27–42.

Lev-Yadun, S., Gopher, A. and Abbo, S. 2000. The cradle of agriculture. *Science Perspective Archaeology* 288(5471), 1602–1603.

Lewis, L.A. and McCourt, R.A. 2004. Green algae and the origin of land plants. *American Journal of Botany* 91(10), 1535–1556.

Lindenmayer, D.B., Fischer, J., Felton, A., Crane, M., Michael, D., Macgregor, C., Montague-Drake, R., Manning, A., and Hobbs, R.J. 2008. Novel ecosystems resulting from landscape transformation create dilemmas for modern conservation practice. *Conservation Letters* 1, 129–135.

Lobell, D.B., Burke, M.B., Tebaldi, C., Mastrandrea, M.D., Falcon, W.P. and Naylor, R.L. 2008. Prioritizing climate change adaption needs for food security in 2030. *Science* 319(5863), 607–610.

Lockie, S., Lyons, K., Lawrence, G. and Haplin, D. 2006. *Going Organic: Mobilizing Networks for Environmentally Responsible Food Production.* Oxfordshire: CAB International.

Longhurst, R. 2001 *Bodies: Exploring Fluid Boundaries.* London: Routledge.

Lorimer, H. 2006. Herding memories of humans and animals. *Environment and Planning D: Society and Space* 24, 497–518.

Lorimer, J. 2010. Elephants as companion species: The 'lively biogeographies' of Asian elephant conservation in Sri Lanka. *Transactions of the Institute of British Geographers* NS 35, 491–506.

Lorimer, J. and Davies, G. 2010. When species meet. Interdisciplinary conversations on interspecies encounters. *Environment and Planning D: Society and Space* 28, 32–33.

Lowe, T.D. and Lorenzoni, I. 2007. Danger is all around: Eliciting expert perceptions for managing climate change through a mental models approach. *Global Environmental Change* 17(1), 131–146.

Lucas, D. and Russell-Smith, J. 1993. *Traditional Resources of the South Alligator Floodplain: Utilisation and Management*, Final consultancy report to the Australian Nature Conservation Agency.

Ludwig, F. and Asseng, S. 2006. Climate change impacts on wheat production in a Mediterranean environment in Western Australia. *Agricultural Systems* 90(1–3), 159–179.

Lulka, D. 2009. The residual humanism of hybridity: Retaining a sense of the earth. *Transactions of the Institute of British Geographers* 34(3), 378–393.

Lunt, D.J., Ross, I.R., Hopley P.J. and Valdes, P.J. 2007. Modelling Late Oligocene C4 grasses and climate. *Palaeogeography, Palaeoclimatology, Palaeoecology* 251(2), 239–253.

Lupton, F.G.H. 1987. History of Wheat Breeding, in *Wheat Breeding. Its Scientific Basis*, edited by F.G.H. Lupton. London: Chapman and Hall, 51–70.

Macindoe, S. L. 1975. History of production of wheat in Australia. *Australian Field Crops, Volume 1 Wheat and Other Temperate Cereals*, edited by A. Lazenby and E.M. Matheson. Sydney: Angus and Robertson, 99–121.

Mackey, B. 2008. Boundaries, data and conservation. *Journal of Biogeography* 35(3), 392–3.

(MAFF) Australian Government Minister for Agriculture Fisheries and Forestry. 2009. *Drought support continues for farming families*, The Hon. Tony Burke MP Media Release DAFF09/243B. Available at http://www.maff.gov.au/media/media_releases/2009/may/drought_support_continues [accessed: 14 October 2009].

Main, G. 2005. *Heartland. The Regeneration of Rural Place*. Sydney: UNSW Press.

Maki, M. and Collin, P. 1997. Coeliac disease. *The Lancet* 349.9057.

Malihot W.C. and Patton, J.C. 1988. Criteria of flour quality, in *Wheat: Chemistry and Technology*, edited by Y. Pomeranz, St. Paul, Minnesota: American Association of Cereal Chemists, Inc, 69–90.

Manning, A.D., Fischer, J., Felton, A., Newell, B., Steffen, W. and Lindenmayer, D.B. 2009. Landscape fluidity – A unifying perspective for understanding and adapting to global change. *Journal of Biogeography* 36(2), 193–199.

Margulis, L. and Schwartz, K. 2009. *Kingdoms and Domains. An Illustrated Guide to the Phyla of Life on Earth*. New York: W. H. Freeman and Co.

Marsden, T. 2000. Food matters and the matter of food: Towards a new food governance? *Sociologia Ruralis* 40(1), 20–29.

Marston, S.A. 2000. The social construction of scale. *Progress in Human Geography* 24(2), 219–42.

Massey, D. 2004. Geographies of responsibility. *Geografiska Annaler* 86(1), 5–18.

McCann, J. 2005. History and memory in Australia's wheatlands, in *Struggle Country: The Rural Ideal in Twentieth Century Australia*, edited by G. Davison and M. Brodie. Clayton: Monash University ePress, 1–17.

McGrath, A. 1987 *Born in the Cattle: Aborigines in Cattle Country*. Sydney: Allen and Unwin.

McGuirk, P.M. 1997. Multiscaled interpretations of urban change: The federal, the state, and the local in the Western Area Strategy of Adelaide. *Environment and Planning D: Society and Space* 15(4), 481–498.

Meinig, D.W. 1962. *On The Margins of the Good Earth: The South Australian Wheat Frontier, 1869–1884*. Adelaide: Rigby Limited.

Mekonnen, M.M. and Hoekstra, A.Y. 2010. A global and high-resolution assessment of the green, blue and grey water footprint of wheat. *Hydrology and Earth System Sciences* 14(7), 1259–1276.

Mercader, J. 2009. Mozambican grass seed consumption during the middle stone age. *Science* 326(5960), 1680–1683.

Mitchell, T.L. 1848. *Journal of an Expedition into the Interior of Tropical Australia in Search of a Route from Sydney to the Gulf of Carpentaria.* London: Longmans.

Mitchell, T. 2002. *Rule of Experts. Egypt, Techno-Politics, Modernity.* Berkeley and Los Angeles: University of California Press.

Mitchell, B.R. (ed.) 2003. *International Historical Statistics – Africa, Asia and Oceania 1750–2000*, 4th Edition. Palgrave Macmillan: New York.

Mitchelle, D.O. and Milke, M. 2005. Wheat: The global market, policies and priorities, in *Global Agricultural Trade and Developing Countries*, edited by M. Ataman and J.C. Beghin. Washington, D.C: The World Bank, 195–214.

Mol, A. 2002. *The Body Multiple: Ontology in Medical Practice.* Durham, NC: Duke University Press.

Morgan, D. 1980. *Merchants of Grain.* New York: Penguin Books.

Morgan, K., Marsden, T. and Murdoch, J. 2006. *World of Food. Place, Power, and Provenance in the Food Chain.* Oxford: Oxford University Press.

Moschini, G. and Hennessy, D.A. 2001. Uncertainty, risk aversion, and risk management for agricultural producers. *Handbook for Agricultural Economics* 1(A), 88–153.

Mosko, M.S. 2009. The fractal yam: Botanical imagery and human agency in the Trobriands. *Journal of the Royal Anthropological Institute* NS 15, 679–700.

Morrison, L.A. 2001. The Percival Herbarium and wheat taxonomy: Yesterday, today and tomorrow, in *Wheat Taxonomy: The Legacy of John Percival,* edited by P.D.S. Caligari and P.E. Brandham. London: Linnean Society, Linnean Special Issue 3, 65–80.

Muehlenchemie (n.d.) Wheat: More than just a plant. Available at http://www.mehlstandardisierung.de/downloads-future-of-flour/FoF_Kap_02.pdf [accessed: 20 October 2011].

Mulvaney, J. and Kamminga, J. 1999. *Prehistory of Australia.* Sydney: Allen & Unwin.

Narayanaswamy, V., Altham, J., Van Berkel, R. and McGregor, M. 2004. Environmental Life Cycle Assessment (LCA) Case Studies for Western Australian Grain Products. Perth: Curtin University. Available at http://www.tud.ttu.ee/material/piirimae/LCA/Case%20studies/LCA%20grain%20products.pdf [accessed: 16 October 2011].

Nelson, G.C., Rosegrant, M.W., Koo, J., Robertson, R., Sulser, T., Zhu, T., Ringler, C., Msangi, S., Palazzo, A., Magalhaes, M., Valmonte-Santos, R., Ewing, M. and Lee, D. 2009. *Climate Change. Impact on Agriculture and Costs of Adaptation.* Washington, DC: International Food Policy Research Institute.

Nesbitt, M. 2001. Wheat evolution: Integrating archaeological and biological evidence, in *Wheat Taxonomy: The Legacy of John Percival,* edited by P.D.S. Caligari and P.E. Brandham. London: Linnean Society, Linnean Special Issue, 37–59.

Niggemann, M., Jetzkowitz, J., Brunzel, S., Wichmann, M.C. and Bialozyt, R. 2009. Distribution patterns of plants explained by human movement behaviour. *Ecological Modelling* 220(9–10), 1339–1346.

O'Brien, L. 1982. Victorian wheat yield trends 1898–1977. *Journal of the Institute of Agricultural Sciences* 48, 163–168.

O'Brien, L., Morell, M., Wrigley, C. and Appels, R. 2001. Genetic Pool of Australian Wheats, in *The World Wheat Book. A History of Wheat Breeding*, edited by A.P. Bonjean and J.F. Angus. Paris: Lavoisier, 611–645.

O'Brien, K., Eriksen, S., Sygna, L. and Naess, L.O. 2006. Questioning complacency: Climate change impacts, vulnerability, and adaptation in Norway. *Ambio* 35(2), 50–56.

O'Connell, J.F. in Torrence, R. 2006. Starch in Archaeology, in *Ancient Starch Research*, edited by R. Torrence and H. Barton. California: Left Coast Press, 20–21.

Olmstead, A.L. and Rhode P.W. 2007. Biological globalization: The other grain invasion. *The New Comparative Economic History. Essays in Honor of Jeffery G. Williamson*, edited by T.J. Hatton, K.H. O'Rourke and A.M. Taylor. Cambridge, MA: The MIT Press, 115–140.

Ortiz, R., Sayre, K.D., Govaerts, B., Gupta, R., Subbarao, G.V., Ban, T., Hodson, D., Dixon, J.M., Ortiz-Monasterio, J.I. and Reynolds, M. 2008. Climate change: Can wheat beat the heat? *Agriculture, Ecosystems and Environment* 126(1–2), 46–58.

Overington, C. 2007. *Kickback: Inside the Australian Wheat Board Scandal.* Crows Nest, NSW: Allen & Unwin.

Pawson, E. 2008. Plants, mobilities and landscapes: Environmental histories of botanical exchange. *Geography Compass* 2/5, 1464–1477.

(PC) Productivity Commission. 2010. Wheat Export Marketing Arrangements. Report No. 51. Canberra: Australian Government Publishers.

Penker, M. 2006. Mapping and measuring the ecological embeddedness of food supply chains. *Geoforum* 37(3), 368–379.

Perkins, J.H. 1997. *Geopolitics and the Green Revolution. Wheat, Genes, and the Cold War.* New York: Oxford University Press.

Perkins, J.H. and Jamison, R. 2008. History, ethics, and intensification in agriculture, in *The Ethics of Intensification: Agricultural Development and Cultural Change*, edited by M. Korthals. Dordrecht, London: Springer, 59–84.

(PFIAA) Pet Food Industry Association of Australia. 2007. *Code of practice for the manufacturing and marketing of pet food.* Available at http://www.pfiaa.com.au/site/files/ul/data_text30/65520.pdf [accessed: 16 October 2011].

Phillips, C. (forthcoming) *Seed Matters: Seed savings practices and politics.* Aldershot: Ashgate.

Pickett, S.T.A., Cadenasso, M.L., Grove, J.M., Boone, C.G., Groffman, P.M., Irwin, E., Kaushal, S.S., Marshall, V., McGrath, B.P., Nilon, C.H., Pouyat, R.V., Szlavecz, K., Troy, A. and Warren, P. 2011. Urban ecological systems:

Scientific foundations and a decade of progress. *Journal of Environmental Management* 92, 331–362.

Piesse, J. and Thirtle, C. 2009. Three bubbles and a panic: An explanatory review of recent food commodity price events. *Food Policy* 34(2), 119–129.

Pimentel, D. and Pimentel, M.H. 2008. *Food, Energy, and Society,* 3rd Edition. Boca Raton: CRC Press.

Piperno, D.R., Weiss, E., Hoist, I. and Nadel, D. 2004. Processing of wild cereal grains in the Upper Palaeolithic revealed by starch grain analysis. *Nature* 430, 670–673.

(PIRSA) Primary Industries and Resources South Australia. 2011. *A History of Agriculture in South Australia.* Available at http://www.pir.sa.gov.au/aghistory/ cereals__and__grains/wheat/wheat_quality/the_early_history_-_wj_farrers_ work [accessed: 20 August 2011].

Plumwood, V. 1993. *Feminism and the Mastery of Nature.* London and New York: Routledge.

Plumwood, V. 2008. Shadow places and the politics of dwelling. *Australian Humanities Review* 44, 139–150.

Potgieter, A.B., Hammer, G.L. and Butler, D. 2002. Spatial and Temporal Patterns in Australian Wheat Yield and their Relationship with ENSO. *Australian Journal of Agricultural Research* 53(1), 77–89.

(PMSEIC) The Prime Minister's Science, Engineering and Innovation Council. 2010. *Australia and Food Security in a Changing World.* Canberra: Australian Government.

Pomeranz, Y. (ed.). 1988. *Wheat: Chemistry and Technology.* St. Paul, Minnesota: American Association of Cereal Chemists, Inc.

Power, E. 2005. Human-nature relations in suburban gardens. *Australian Geographer* 36(1), 39–53.

Probyn, E. 1993. *Sexing the Self. Gendered Positions in Cultural Studies.* London: Routledge.

Probyn, E. 2000. *Carnal Appetites. Food Sex Identities.* New York: Routledge.

Qui, Y., Lee, J., Bernasconi-Quadroni, F., Soltis, D., Soltis, P., Zanis, M., Zimmer, E., Chen, Z., Savolainen, V., and Chase, M. 1999.The earliest angiosperms: Evidence from mitochondrial, plastid and nuclear genomes. *Letters to Nature* 401, 404–407.

Raven, J.A. and Allen, J.F. 2003. Genomics and chloroplast evolution: What did cyanobacteria do for plants? Mini review. *Genome Biology* 4(3), 209.

Reid, S., Smith, B., Caldwell, W. and Belliveau, S. 2007. Vulnerability and adaptation to climate risks in Ontario agriculture. *Mitigation and Adaptation Strategies for Global Change* 12(4), 609–637.

Reisner, M. 1986. *Cadillac Desert: The American West and Its Disappearing Water.* New York. NY: Viking Press.

Reynolds, M. and Borlaug, N.E. 2006. Applying innovations and new technologies for international collaborative wheat improvement. *Journal of Agricultural Science* 144(2), 95–110.

Rival, L. 1993. The growth of family trees: Understanding Huaorani perceptions of the forest. *Man* 28(4), 635–652.

Rival, L. (ed.). 1998. *The Social Life of Trees: Anthropological Perspectives on Tree Symbolism.* Oxford: Berg.

Rival, L. 2006. Amazonian historical ecologies. *Journal of the Royal Anthropological Institute* (N.S.) 12, S79-S94.

Robbins, P. 2007. *Lawn People.* Philadelphia: Temple University Press.

Robinson, M.E. 1976. The New South Wales Wheat Frontier 1851–1911. Department of Human Geography, Research School of Pacific Studies, The Australian National University, Canberra. PhD Thesis

Robinson, J.B., Freebain, D.M., Dimea, J.P., Dalal, R.C., Thomas, G.A. and Weston E.J. 1999. Modelling wheat production from low-rainfall farming systems in northern Australia. *Environment International* 25(6/7), 861–870.

Rodaway, P. 1994. *Sensuous Geographies. Body, Sense and Place.* London: Routledge.

Roe, E.J. 2006a. Things becoming food and the embodied, material practices of an organic food consumer. *Sociologia Ruralis* 46, 104–121.

Roe, E.J. 2006b. Material connectivity, the immaterial and the aesthetic of eating practices: An argument for how genetically modified foodstuff becomes inedible. *Environment and Planning A* 38, 465–481.

Roggeveen, K. 2010. Tomato journeys from farm to fruit shop. MSc thesis. University of Wollongong.

Rose, D.B. 1996. *Nourishing Terrains. Australian Aboriginal Views of Landscape and Wilderness.* Canberra: Australian Heritage Commission.

Rost, T.L., Barbour, M.G., Stocking, C.R. and Murphy, T.M. 2006. *Plant Biology,* 2nd Edition. Southbank, Vic: Thompson Brooks Cole.

Ruddiman, W.F. 2003. The anthropogenic greenhouse era began thousands of years ago. *Climatic Change* 61(3), 261–293.

Sagoff, M. 2003. The plaza and the pendulum: Two concepts of ecological science. *Biology and Philosophy* 18(4), 529–552.

Salamini, F., Ozkan, H., Brandolini, A., Schafer-Pregl, R. and Martin, W. 2002. Genetics and geography of wild cereal domestication in the near east. *Nature Reviews Genetics* 3 (June), 429–441.

Saltzman, K., Head, L. and Stenseke, M. 2011. Do cows belong in nature? The cultural basis of agriculture in Sweden and Australia. *Journal of Rural Studies* 27, 54–62.

Sauer, C. 1952. *Agricultural Origins and Dispersals.* New York: American Geographical Society.

Saunders, C., Barber, A. and Taylor, G. 2006. *Food Miles – Comparative Energy/Emissions Performance of New Zealand's Agriculture Industry,* Research Report 285, Agribusiness & Economics Research Unit Lincoln University, Lincoln.

Sears, E.R. 1981. Transfer of alien genetic material to wheat, in *Wheat Science – Today and Tomorrow*, edited by L.T. Evans and W.J. Peacock. Cambridge: Cambridge University Press, 75–89.

Sheller, M. and Urry, J. 2006. The new mobilities paradigm. *Environment and Planning A* 38, 207–226.

Shepherd, S. and Gibson, P.R. 2006. Understanding the gluten-free diet for teaching in Australia. *Nutrition & Dietetics*, 63, 155–165.

Shiva, V. 1993. *The Violence of the Green Revolution. Third World Agriculture, Ecology and Politics*. London: Zed Books Ltd.

Silburn, D.M., Freebairn, D.M., and Rattray, D.J. 2007. Tillage and the environment in sub-tropical Australia – Tradeoffs and challenges. *Soil and Tillage Research* 97(2), 306–317.

Simpson, M.G. 2006. *Plant Systematics*. Amsterdam: Elsevier Academic Press.

Singh, G. 1999. *Plant Systematics*. Enfield USA: Science Publishers.

Slovic, P., Finucane, M.L., Peters, E. and MacGregor, D.G. 2004. Risk as analysis and risk as feelings: Some thoughts about affect, reason, risk and rationality. *Risk Analysis* 24(2), 311–322.

Smit, B. and Skinner, M.W. 2002. Adaptation options in agriculture to climate change; a typology. *Mitigation and Adaptation Strategies for Global Change* 7(1), 85–114.

Staddon, C. 2009a. Towards a critical political ecology of human-forest interactions: Collecting herbs and mushrooms in a Bulgarian locality. *Transactions of the Institute of British Geographers* 34(2), 161–176.

Staddon, C. 2009b. The complicity of trees: The socio natural field of/for tree theft in Bulgaria. *Slavic Review* 68, 70–94.

Stephens, D.J. and Lyons, T.L. 1998. Rainfall-yield relationships across the Australian wheatbelt. *Australian Journal of Agricultural Research* 49, 211–223.

Stevenson, A. (ed.). 2010. *Oxford Dictionary of English*. Oxford: Oxford University Press. Available at http://www.oxfordreference.com/views [accessed: 19 August 2011].

Strathern, M. 1996. Cutting the Network. *Journal of the Royal Anthropological Institute* NS 2, 517–535.

Sturt, C. 1849. *Narrative of an expedition into Central Australia: performed under the authority of Her Majesty's Government, during the years 1844, 5, and 6, together with a notice of the Province of South Australia, in 1847*. Adelaide: Corkwood Press.

Tanno, K. and Willcox, G. 2006. How fast was wild wheat domesticated? *Science* 311 (5769), 1886.

Terashima, H. 2001. The relationships among plants, animals, and man in the African tropical rain forest. *African Study Monographs* 27, 43–60.

Terrell, J.E., Hart, J.P., Barut, S., Cellinese, N., Curet, A., Denham, T.P., Haines, H., Kusimba, C.M., Latinis, K., Oka, R., Palka, R., Pohl, M.E.D., Pope, K.O., Staller, J.E. and Williams, P.R. 2003. Domesticated landscapes: The subsistence ecology of plant and animal domestication. *Journal of Archaeological Method and Theory* 10(4), 323–368.

Terrell, J. 2006. Human biogeography: Evidence of our place in nature. *Journal of Biogeography* 33(12), 2088–2098.

Thanos, C.A. 1994. Aristotle and Theophrastus on plant-animal interactions, in *Plant Animal Interactions in Mediterranean-Type Ecosystems*, edited by M. Arianoutsou and R.H. Groves. Netherlands: Kluwer Academic Publishers, 3–11.

Thomas, W.L. (ed.). 1956. *Man's Role in Changing the Face of the Earth*. Chicago: University of Chicago Press.

Thomas, J., 1991. *Rethinking the Neolithic*. Cambridge: Cambridge University Press.

Thornton, P.K., Jones, P.G., Alagarswamy, G. and Andresen, J. 2008. Spatial variation of crop yield response to climate change in East Africa. *Global Environmental Change* 19(1), 54–65.

Tindale, N.B. 1974. *Aboriginal Tribes of Australia*. Berkeley: UCLA Press.

Tindale, N.B. 1977. Adaptive significance of the Panara or grass seed culture of Australia, in *Stone Tools as Cultural Markers: Change, Evolution and Complexity*, edited by R.V.S. Wright. Canberra: Australian Institute of Aboriginal Studies, 345–9.

Toothill, E. (ed.). 1984. *The Penguin Dictionary of Botany*. London: Penguin Books.

Trethowan, R.M., Reynolds, M.P., Ortiz-Monasterio, J.I. and Ortiz, R. 2007. The Genetic Basis of the Green Revolution in Wheat production, in *Plant Breeding Reviews*, edited by I.L. Goldman, C.H. Michler and R. Ortiz. New Jersey: John Wiley and Sons Inc. 28, 39–58.

Trewavas, A. 2005. Green plants as intelligent organisms. *TRENDS in Plant Science* 10(9), 413–419.

Trewavas, A. 2002. Mindless mastery. *Nature Concepts* 415, 841.

Trigger, D.S. 2008. Indigeniety, ferality, and what 'belongs' in the Australian bush: Aboriginal responses to 'introduced' animals and plants in a settler-descendant society. *Journal of the Royal Anthropological Institute* (N.S.) 14, 628–646.

Truett, J.C. 2010. *Grass in Search of Human Habitat*. California: UCA Press.

Tsing, A.L. 2005. *Friction: An Ethnography of Global Connection*. Princeton, NJ: Princeton University Press.

Tubiello, F.N. and Fischer, G. 2007. Reducing climate change impacts on agriculture: Global and regional effects of mitigation, 2000–2080. *Technological Forecasting and Social Change* 74(7), 1030–1056.

Tzvelev N.N. 1989. The system of grasses (Poaceae) and their evolution. *The Botanical Review* 55(3), 141–204.

Ueda, M., Nakamura, Y. and Okada, M. 2007. Endogenous factors involved in the regulation of movement and 'memory' in plants. *Pure and Applied Chemistry* 79(4), 519–527.

Ueda, M., Tokunaga, T., Okada, M., Nakamura, Y., Takada, N., Suzuki, R. and Kondo, K. 2010. Trap-closing chemical factors of the venus flytrap *Dionaea muscipulla* Ellis. *ChemBioChem* 11(17), 2378–2383.

(USDA) United States Department of Agriculture. 2011. Production, Supply and Distribution Online database (PSDOnline). Available at: http://www.fas.usda. gov/psdonline/psdHome.aspx [accessed: 24 August 2011].

Valentine, G. 2008. Living with difference: Reflections on geographies of encounter. *Progress in Human Geography* 32, 323–337.

Vrydaghs, L. and Denham, T. 2007. Rethinking agriculture: Introductory thoughts, in *Rethinking Agriculture. Archaeological and Ethnoarchaeological Perspectives*, edited by T. Denham, J. Iriarte and L. Vrydaghs. California: Left Coast Press, 1–15.

Wackernagel, M. and Rees, W. 1996. *Our Ecological Footprint: Reducing Human Impact on the Earth*. Gabriola Island, BC, Canada: New Society Publishers.

Waitt, G., Caputi, P., Gibson, C., Farbotko, C., Head, L., Gill, N. and Stanes, E. 2012. Sustainable Household Capability: Which households are doing the work of environmental sustainability? *Australian Geographer* 43(1), 51-74.

Wang, E., Xu, J., Jiang, Q. and Austin, J. 2009. Assessing the spatial impact of climate on wheat productivity and the potential value of climate forecasts at a regional level. *Theoretical and Applied Climatology* 95(3), 311–330.

Wang, J., Wang, E., Liu, D. 2011. Modelling the impacts of climate change on wheat yield and field water balance over the Murray–Darling Basin in Australia. *Theoretical and Applied Climatology* 104(3–4), 285–300.

Warren, J.F. 1969. Australian wheat yield trends. *Journal of the Institute of Agricultural Sciences* 35, 240–252.

Waterhouse, W.L. 1936. Some observations on cereal rust problems in Australia. *Proceedings of the Linnean Society NSW*, 61, v–xxxviii.

Watson, S. 2006. *City Publics: The (Dis)enchantment of Urban Encounters*. London: Routledge.

Watts, M. 2005. Commodities, in *Introducing Human Geographies*, edited by P. Cloke, P. Crang and M. Goodwin. Abingdon, UK: Hodder Arnold, 527–546.

Wei, X., Conway, D., Erda, L., Yinlong, W., Hui, J., Jinhe, J., Holman, I. and Yan, L. 2009. Future cereal production in China: The interaction of climate change, water availability and socio-economic scenarios. *Global Environmental Change* 19(1), 34–44.

Whatmore, S. 2002. *Hybrid Geographies. Natures Cultures Spaces*. London: Sage.

Whatmore, S. and Thorne, L. 1997. Nourishing Networks: Alternative Geographies of Food, in *Globalising Food: Agrarian Questions and Global Restructuring*, edited by D. Goodman, and M.J. Watts. London: Routledge, 287–304.

Whistler, R.L. and Daniel, J.D. 2000. Starch, in *Kirk-Othmer Encyclopedia of Chemical Technology*.

Whitwell, G., Sydenham, D. and The Australian Wheat Board. 1991. *A Shared Harvest. The Australian Wheat Industry, 1939–1989*. Melbourne: Macmillan Education Australia Pty Ltd.

Williams, J. 2001. Towards sustainable land management. *Food for healthy people and a healthy planet*. Internet conference, September 2001. www.natsoc.org. au. Accessed by G. Main, February 2002.

Wolch, J. 2007. Green urban worlds. *Annals of the Association of American Geographers* 97(2), 373–84.

Worster, D. 1979. *Dust Bowl: The Southern Plains in the 1930s*. New York: Oxford University Press.

Worster, D. 1985. *Rivers of Empire: Water, Aridity, and the Growth of the American West*. New York: Pantheon.

Wrangham, R.W., Jones, J.H., Laden, G., Pilbeam, D. and Conklin-Brittain, N. 1999. The raw and the stolen: Cooking and the ecology of human origins. *Current Anthropology* 40(5), 567–594.

Wrigley, C.W. and Rathjen, A. 1981. Wheat breeding in Australia, in *Plants and Man in Australia*, edited by D.J. Carr and S.G.M. Carr. Sydney: Academic Press, 96–135.

Young, A. 2000. *Environmental Change in Australia since 1788*. Melbourne: Oxford University Press.

Xiao, S., Zhang, Y. and Knoll, A.H. 1998. Three dimensional preservation of algae and animal embryos in a neoproterozoic phosphorate. *Nature* 391(6667), 553–558.

Zeder, M.A. 2006. Central questions in the domestication of plants and animals. *Evolutionary Anthropology* 15(3), 105–117.

Ziervogel, G., Bharwani, S. and Downing, T.E. 2006. Adapting to climate variability: Pumpkins, people and policy. *Natural Resources Forum* 30(4), 294–305.

Zohary, D. and Hopf, M. 2000. *Domestication of Plants in the Old World. The Origin and Spread of Cultivated Plants in West Asia, Europe and the Nile Valley*. Oxford: Oxford University Press.

Index